W9-CIF-004

THE OXFORD BOOK OF AUSTRALIAN RELIGIOUS VERSE

For my father

THE OXFORD BOOK OF AUSTRALIAN RELIGIOUS VERSE

edited by KEVIN HART

Melbourne

OXFORD UNIVERSITY PRESS

Oxford Auckland New York

OXFORD UNIVERSITY PRESS AUSTRALIA

Oxford New York Toronto
Delhi Bombay Calcutta Madras Karachi
Kuala Lumpur Singapore Hong Kong Tokyo
Nairobi Dar es Salaam Cape Town
Melbourne Auckland Madrid
and associated companies in
Berlin Ibadan

OXFORD is a trade mark of Oxford University Press

First published 1994

National Library of Australia
Cataloguing-in-Publication data:

The Oxford book of Australian religious verse.

Includes index.
ISBN 0 19 553498 0.

1. Religious poetry, Australian. I. Hart, Kevin, 1954–
II. Title: Book of Australian religious verse.

A821.0080382

This project has been assisted by the Commonwealth Government through the Australia Council, its arts funding and advisory body.

Cover design by Guy Mirabella
Typeset by Solo Typesetting, South Australia
Printed in Australia by Brown Prior Anderson Pty Ltd
Published by Oxford University Press,
253 Normanby Road, South Melbourne, Australia

Contents

INTRODUCTION

Australian religious poetry: what does this expression mean? At first glance the answer seems perfectly clear and dry. There exists a region of poetry, and we wish to limit ourselves to a part that can be called religious. Moreover, we are strictly concerned with that portion of religious poetry which comes from Australia. It is a simple answer and to some extent an inevitable one. Yet once we accept it all sorts of difficulties arise. Just how easy is it to delimit a field of poetry, if there is such a thing?

As soon as you propose a definition of poetry, even one that might meet with broad approval in the late twentieth century, there will be borderline cases that cannot be dismissed. If you frame a definition with only canonical verse in mind you may be pained to exclude the charms, jingles and riddles that in all likelihood led you to poetry in the first place. The same gesture will lead you to omit many if not all of the contemporary poems that intrigue and sustain you. Almost certainly you will put aside broadsides, folk songs and popular lyrics—only to find yourself trying to disentangle questions of value from questions of definition. The same severity would leave no room for clerihews, double dactyls or nonsense verse, and that also would be regrettable. If you agree that poetry should not be restricted to verse (whether formal or free), then concrete poetry and prose poetry will bid for places. Having gone that far, what about 'found poems'? I mean snatches from advertisements, catalogues or graffiti that were never intended to have literary merit but which suddenly shine with poetic effects when read in a suitable context. And if you include prose poems someone is sure to point out that there are poetic passages in autobiographies, diaries, novels and stories, almost anywhere in fact. So far we have been thinking only of writing in English. But if you can be persuaded that not all poetry gets lost in translation, then poems originally composed in other languages will demand your attention, giving rise to worries about linguistic and cultural fidelity.[1] Performance poetry presents a tricky case, especially for an anthologist working towards the end of the print era. To be appreciated, oral verse must be presented in person or on audiovisual cassette. And until anthologies are published on compact disks, computer poems written in hypertext must be passed over.

These few examples show how the borders surrounding poetry are broken and shifting, not on one front but on several. Attempts to define poetry by excising all dubious or marginal cases impoverish what they set out to preserve. But what of the alternative, of not worrying much about circumscribing poetry in a Socratic manner and letting its various usages go their ways for as long as they can? Faced with this possibility people sometimes shudder as if being led to the brink of an abyss. There never has been a groundless free-for-all with poetry, though. (Or if there has, or even if there has never been anything but that, it has never upset people as much as they imagine they would be or feel they should be.) The full range of what counts as poetry changes over time and place, and conceptual difficulties emerge most forcefully when one tries to encompass all those historical and cultural differences in the one formulation. The question 'Is this a poem?' can always be asked, and there are moments when it needs to be asked. And yet there are also circumstances when it is not the most urgent or even the most useful question to pose. Sometimes recourse to definition can be a step backwards, an evasion of poetry and a fear of the unknown. In the words of Maurice Blanchot: 'To read a poem is not to read yet another poem.'[2]

Needless to say, 'religious' is at least as hard to pin down as 'poetry'. The word can be defined this way or that, sociologically or theologically, while admitting that in practice these categories overlap substantially and cannot always be separated even under the glare of theory. From atheists and the pious alike, there is always pressure to restrict 'religious' to a traditional devotion; after all, in one sense the word has long come to mean 'strict'. Rebelling against this curbing of meaning (while acknowledging that religion can be most intense when most austere, long before it flares into enthusiasm or fanaticism, themselves important religious phenomena), one can readily be enticed to take the exact opposite direction. Yet once we begin to extend the scope of 'religious' it is hard to know when and how to stop. The difficulties are bad enough when dealing with one faith, let alone when attempting to keep several in play. Certainly we have to accept natural and revealed religions and certainly we have to go well past institutional borders. Many people for whom churches and temples mean little or nothing acknowledge a higher power—personal or impersonal, singular or plural, transcendent or immanent—which they take to influence or even guide their lives. Once outside organised religion can we also pass beyond the observance of rites? Not if by 'outside' we have in mind only those individuals of a mystical or occultist tendency. But 'rites' can mean more than ceremonies and sacraments; the concept also includes the customs and habits associated with belief. That gives 'rites' a very wide extension, too wide some are bound to say. Yet there will always be an Immanuel Kant to insist that religion is essentially a matter of morality and that works of Grace, miracles, mysteries, and means of Grace are mere supplements to be kept outside religion's true and proper limits. And there will always be people who will side with him.

What of men and women who do not worship a higher power but who remain devoted to a religious teaching? Consider the Taoists, followers of Lao-tzu whose doctrines are gathered in the *Tao-te-ching*. When Taoists talk of the *Tao* or Way, they do not have a Supreme Being in mind, perhaps not even Being as usually understood in the West. Certainly it is nothing that could be regarded as God or a god. Even at their most speculative, when claiming that *Tao* is inaudible, invisible, unchangeable and unnameable, they are using negations to indicate a transcendental principle, not a transcendent deity. Like the lines dividing religion from mythology, the borders between religion and philosophy tend to be wavy and interrupted. In both situations the limits are apparent not only at the edges, giving rise to fiddly cases; they are often to be found at the doctrinal centre of a religion. Every time a Catholic recites the Nicene Creed at Mass he or she assents, at least tacitly, to certain notions of Aristotelian metaphysics. (By the same token philosophy can be deeply marked by myth—Plato's fable of the cave and Nietzsche's invocation of the eternal return, for example—or, equally, philosophy can be touched by an intense religious longing: Plotinus's *Enneads* and Hegel's *Phenomenology of Spirit* are prime examples.) That said, in the Western world Taoism along with Confucianism is usually treated far more fully in encyclopaedias of religion than philosophy.

Not so with Marxism; and yet in both East and West it has given rise to the most widespread and powerful eschatological movement of the century. When people talked in the 1950s and 1960s of 'the Communist Church' in the former Soviet Union, the expression may have been intended polemically, linking the bureaucratic machinations of Moscow and the Vatican, but it brought and still brings to the fore some ways in which communism reset and reworked Christian notions of community and along with them the theological virtues of faith, hope and charity. In the first decades of this century Marxism promised egalitarianism, and the glow of this promise long outlived the economic and political performances of successive governments. In recent decades it became increasingly apparent even to deeply committed socialists that faith, hope and charity cannot be realised within the horizon of dialectical materialism. For all that, there can be no simple story of communism versus religion, not least of all because Christians, Jews and Muslims (I speak broadly and only of the West) have suffered as harshly in our time under anti-communist regimes. Besides, both ideological extremes are perpetually traversed by religious motifs, while most religious groups have keen interests in both the political at large as well as in party politics. And there is another, more harrowing reason not to single out socialism. From the 1917 Russian Revolution to the dismantling of the Berlin Wall and beyond, numerous men, women and children of all faiths have fallen victim to people of other faiths or of the same belief—and all in the name of God. Inhumanity is not restricted to the secular world.

Having come this far, should we include atheism as a religious phenomenon? We have no choice but to do so, providing we distinguish atheisms that take religious questions seriously from those that do not. The view that God does not exist, that all deities are illusions, even pernicious ones, is often prosecuted with a fervour that derives its energy and sometimes its style from the faith it refuses. There are fundamentalists and evangelicals in atheism as surely as there are in theism. One of the most rhetorically effective atheists of our era is Friedrich Nietzsche. When announcing his terrible message 'God is dead', he assumed the character of a madman who 'ran to the market place, and cried incessantly: "I seek God! I seek God!"' before delivering his desolating sermon:

> 'Whither is God?' he cried; 'I will tell you. *We have killed him*—you and I. All of us are his murderers. But how did we do this? How could we drink up the sea? Who gave us the sponge to wipe away the entire horizon? What were we doing when we unchained this earth from its sun? Whither is it moving now? Whither are we moving? Away from all suns? Are we not plunging continually? Backward, sideward, forward, in all directions? Is there still any up or down? Are we not straying as through an infinite nothing? Do we not feel the breath of empty space? Has it not become colder? Is not night continually closing in on us? Do we not need to light lanterns in the morning? Do we hear nothing as yet of the noise of the gravediggers who are burying God? Do we smell nothing as yet of the divine decomposition? Gods, too, decompose. God is dead. God remains dead. And we have killed him.[3]

Nietzsche would concede that there are many for whom God remains an abiding presence (although he would also maintain that, like the philosophers, they mistakenly infer the nature of reality from contingent grammatical categories). His central point is that believers and non-believers alike have not yet admitted to themselves that the world is no longer turned decisively toward God, that the realm of the transcendent has lost its hold on how we live and how we die.

If the divine can fade from the world it can always return, possibly in a very different form. As modern thinkers and writers have discovered time and again, it is easier to dismiss the word and concept 'God' than to eliminate the category of the sacred. Without slighting the illuminations of teachers of all faiths, I have to say that it happens—and certainly with great force in our own era—that some of the most religiously acute observations are made by those who cannot in all conscience avow God. Doubt is not always directly opposed to faith; it can be a twisting path to a faith; and everyone who grows in belief must risk asking 'What does "God" mean?', 'How can I speak of God?' and 'Where is God?'—and not just in moments of abjection and suspicion but also in times of contentment and elation. Amongst the twentieth century's most intense religious writing I would include passages from Samuel Beckett's *Waiting for Godot*, Martin Heidegger's meditations on Hölderlin, Franz Kafka's *The Castle* and Simone Weil's *Notebooks*. These texts do not converge at any one point but taken together they urge us not to use the word 'God' lightly. They teach that to pronounce the divine name too quickly, or to circumscribe it too tightly, can crush a question rather than answer it. And they suggest that God can be revealed more truly in the question than in the answer. Thinking along these lines, who can brood on Edmond Jabès's *The Book of Questions* without gaining a deeper understanding of the anguished attempts to fathom God after the Holocaust? If nothing else, you will be haunted by one or more of the fragments uttered by his imaginary rabbis: 'God is a questioning of God' one of them hazards, and then—almost as a long deferred reply—another ventures, 'God is always in search of God'.[4] Strange words, though not to a Kabbalist or a Rhineland mystic. In an interview Jabès says, 'What I mean by God in my work is something we come up against, an abyss, a void, something against which we are powerless. It is a distance . . . the distance that is always between things . . . We get to where we are going, and then there is still this distance to cover. And a moment comes when you can no longer cover the distance; you get there and you say to yourself, it's finished, there are no more words. God is perhaps a word without words.'[5]

Let us pause for a moment and take stock. When the words 'poetry' and 'religion' are allowed to overlap, like two fuzzy Venn diagrams, there forms a rich and unruly set that includes everything and anything from carols, hymns and prayers to miracle plays and verse essays in doctrine, to stanzas that affirm what Friedrich Schleiermacher called a 'feeling of absolute dependence' or what Romain Rolland dubbed an 'oceanic feeling', to assertions of God's death and quieter confessions of a gradual fading of belief, and a great deal else besides. There is no shortage of ways to tidy up this sprawl. Some are drastic. You can say with Karl Barth that Christianity is not a religion because religion is a human attempt to find God while Christianity is the story of God going in search of sinful men and women. That definition eliminates the category of Christian religious verse altogether, though not religious verse as such. But to see that being done we have only to turn to Samuel Johnson's authoritative comments in his *Life of Waller* (1781):

> Contemplative piety, or the intercourse between God and the human soul, cannot be poetical. Man admitted to implore the mercy of his Creator and plead the merits of his Redeemer, is already in a higher state than poetry can confer.
>
> The essence of poetry is invention; such invention as, by producing something unexpected, surprises and delights. The topicks of devotion are few, and being few are universally known; but, few as they are, they can be made no more; they can receive no grace from novelty of sentiment, and very little from novelty of expression.[6]

Adopting an opposite stance, you can claim, as very different writers have done over the years, that poetry is Orphic: a transfiguration of nature into song that overcomes death and the grave. And so, regardless of its explicit themes, all poetry is at heart sacred. This certainly expands the category of religious poetry, so much so, though, that in effect it abolishes it no less surely than Johnson does.

Accepting none of these extreme solutions, I would like to keep all I can in this mixed and ragged cluster of texts called 'religious poetry' and overlay one more set, 'Australian'. It will not solve any difficulties but will add some more, in both the fields of poetry and religion. When we put 'Australian' before either 'poetry' or 'religion' it seems as if we have grafted something very new onto something very old. We think of a country that celebrated its bicentennial as recently as 1988, two centuries after being declared *terra nullius* and claimed for the Crown by Captain James Cook. As likely as not, we think of Australian poetry working with, alongside or sometimes against a long tradition of European poetics, and we think of Australian religion as an inflection of European Christianity. This accounts for much of our poetry and our institutional religiosity, though certainly not for the traditional songs and sacred ceremonies of the land's Aboriginal peoples. It must be remembered that the word 'Australian' derives from a seventeenth-century coinage, *terra australis*, that names an uncharted space—Australasia, Polynesia and 'Magellanica'—as viewed from a European perspective. So to call traditional Aboriginal chants Australian is not without problems, for it involves the colonising gaze whose effects have brutalised and endangered those same traditions. And in a similar way, one should pause a while before construing the chants as poetry lest they be measured purely and simply against aesthetic norms that remain largely exterior to them. The song cycles are living performances, allowing great variation of movement, music and words. They cannot be abstracted from their living contexts without enormous losses.

One way to approach Australian poetry and religion is by asking what languages they use. English is culturally predominant. But this should not blind us to the fact that the land has a hundred and fifty or so Aboriginal tongues (and once had many more), and over a hundred others drawn from Asia and Europe. To judge from local anthologies published during the last few decades, 'Australian poetry' has come to mean a canon of writing composed in English by anglophones living in or born in Australia. In recent years especially that canon has been supplemented by translations from Aboriginal song cycles. We know that verse is composed here in German, Greek and Italian—and very occasionally translations from these works appear in anthologies—but at the moment it is impossible to form any impression of poetry being written locally in Arabic, Chinese, Hungarian, Polish, Spanish, Turkish, Ukranian, Vietnamese, or any of the other community languages.

Even if we were to restrict ourselves to material written in English, Australian poetry would not be all of a piece. Taken by themselves, historical markers are unreliable. The year 1788 may signal the start of British colonialisation, but it can hardly indicate the origin of the country's poetry. On the one hand there are writers whose ethnic background sets them at variance with the myth of national origin. The verse of Antigone Kefala and Dimitris Tsaloumas, for example, moves to the rhythms of different historical and mythological times. While on the other hand there are poets from British stock whose main influences are Asian, American or European. Robert Gray's imaginative world, for instance, has been shaped by his reading of Francis Ponge, Charles Reznikoff, and Japanese Zen poets. In much the same way, when taken by themselves geographical markers are not to be trusted. Peter Porter is thought by some to be our finest living poet. He was born in Brisbane, and returns to Australia from time to time; but for the last forty years he has lived in London and made a solid reputation there. Where does the Australia in 'Australian poetry' begin and end?

The brief list of community languages spoken in Australia is enough to suggest that Christianity, though culturally predominant, is far from being the only faith practised in the country. It also indicates that Australian Christianity is not monolithic. Catholic traditions from East and West are represented, as are the various families of Protestantism and other, less easily categorised groups like Baha'i, Christian Science, the Society of Friends, and Unitarians. There are many Aboriginal religions—not, as is commonly supposed, just the one. Of the major world faiths, Buddhism, Hinduism, Islam and Judaism are all to be found, though all in relatively small numbers when set against the population as a whole; and there are tiny numbers of other sects like Rosicrucians, Swedenborgians and

Zoroastrians.[7] The impact of these groups on Australian poetry is very uneven. Although Islam is our second largest religious community—a consequence of steady immigration from Lebanon and Turkey over the past thirty years—one would be hard-pressed to find much Muslim poetry written in or translated into English. The handful of Australian poets who have fallen under the spell of Rumi or Ghalib have done so without the mediation of Islam. Conversely, while statistics tell us that there are few Australians who follow Chinese religions, several local poets with British forebears have been strongly drawn to Buddhism, Taoism, or a synthesis of the two, and not only use themes from those faiths but can be counted as adherents.

Potentially, then, the field of Australian religious poetry is vast and chaotic. When reading with this anthology in view it was often useful, if sometimes daunting, to keep in mind a sense of that possible imbroglio and of those needling fringe cases. It made me spend weeks and months studying small-press and fugitive publications, sifting through fragile collections of nineteenth-century verse, tracking down books and pamphlets by people born in Australia but now living overseas, writing to representatives of religious groups who might know of newsletters that printed poems, as well as rereading familiar poems to see if they showed a radiance I had not noticed before. The idea of a discrete body of work abiding ideally in the space marked out by three overlapping sets was beguiling, and doubtless forced me to look in places I might otherwise have passed by. But after a while I found myself compelled to change my guiding metaphor. I started to think of 'Australia', 'religion' and 'poetry' as different threads that had been interlaced, and I was encouraged when I recalled one of Ludwig Wittgenstein's remarks in the *Philosophical Investigations* explaining the notion of 'family resemblance'. He wrote of how when 'spinning a thread we twist fibre on fibre', and observed that 'the strength of the thread does not reside in the fact that some one fibre runs through its whole length, but in the overlapping of many fibres'.[8]

Poetry, religion and Australia are threads made from the overlapping of various fibres, not one of which runs through a whole length. Reading and rereading for this anthology I found there to be no single knot that ties together all three to make 'Australian religious poetry'. There are several knots and they do not always tie up the same fibres. What makes David Curzon's 'Proverbs 6: 6' an Australian religious poem is very different from what allows us to describe Nigel Roberts's 'Reward/for a missing deity' and the Yirrkalla people of Arnhem Land's 'The Djanggawul Song Cycle' in the same three words. David Curzon was born in Melbourne but has lived in New York for most of his life. His poem is a midrash on a biblical text. Taken from the Hebrew verb *drash*, meaning 'search', midrash is a rigorous yet imaginative inquiry into the substance of scripture with a view to link it to present concerns and questions. In writing this way Curzon joins himself to a tradition of commentary that can be traced back centuries before the common era. Nigel Roberts was born in New Zealand but has lived most of his adult life in Australia. By and large, his writing answers to American rather than Australian verse. 'Reward/for a missing deity' gains its force by twinning irreverent speculation with a sense of God's absence. And the Yirrkalla people's song cycle is, as I suggested earlier, not a poem in a traditional Western sense; it is abstracted from its ceremonial contexts, translated from an Aboriginal tongue and in any case is older than the word 'Australia'.

So there is considerable variety in this anthology. European faiths rub up against home-grown and imported materialisms, and at times both find themselves in dialogue with Asian spiritualities. All the same, there is rather less variety than I imagined there might be. There was no difficulty locating a range of poems that work with or against Christianity understood fairly broadly, and there are a number of Aboriginal song cycles to choose from. Nor was it hard to find material that draws from Buddhism and Taoism. The two longest pieces I have included, Harold Stewart's 'Lingering at the Window of an Inn after Midnight' and Robert Gray's 'Dharma Vehicle', explore Buddhism through the lenses of very different poetics. Randolph Stow appears to have been fascinated by Taoism from early in his writing, while Judith Wright moves from Christian to Taoist motifs in her later writing. Of other major world religions I have found little material to include. In making selections I placed the accent firmly on Australian religious *poetry*. If a text did not work as poetry for me, it did not keep my attention for long. While I found some material that looked to Islam or was inspired by New Age spirituality I could find nothing raised to the level of art, and before I started work I had made it a rule not to include anything simply because it represented a kind of religiosity.

Before I started work I also proposed to set aside poems that did no more than describe churches and graveyards, or that simply presented clerics and churchgoers as 'characters'. In principle it seems easy to separate those poems that are religious from those that brush against religion while taking no real interest in it. In practice, however, it can be hard to sustain that distinction. A poem that sets out to evoke something conventionally associated with religion can get caught up in what it describes. To take one example out of many in this anthology, B. R. Whiting's lyric 'Gandhi, 1946' relates an encounter with the great Hindu leader. In the first stanza the speaker makes it clear that he was not drawn to Gandhi or his disciples, and that he was unimpressed when Gandhi presumed to lean on his arm for strength. The story is told dispassionately, right up to and almost including the closing four lines:

> And now I like to think
> How, after I had escorted him to the car,
> The Khitmugar came up
> To ask permission, if he might touch that arm.

All that we hear (and that indirectly) is a request, 'if he might touch that arm', which lightly evokes the Khitmugar's customary politeness while also insisting that the speaker's arm is no longer simply in the realm of the customary. It has entered the world of the sacred, become 'that arm' for the Khitmugar and, for as long as the memory lasts, for the speaker himself.

As I read and reread it became increasingly plain that those Australian poets most given to formal experimentation tended to be uninterested in matters of the spirit. When I looked to folk songs and popular lyrics, I could find occasional allusions to God but nothing that took either doubt or faith as its main concern. Hymns there were, of course, examples of innovative liturgy set out as free verse, and passages from libretti. In the context of the selection as a whole, they seemed out of their element. With nineteenth-century poetry there was no shortage of religious feeling. Try though I might, I could not find much to include. Our Victorian age had a good deal of characteristic verse but little distinguished poetry. This is one reason why I finally decided not to arrange the anthology chronologically. Also, though, when we are dealing with two unmatchable histories—the ancient Aboriginal traditions and those that follow colonisation—chronology hinders effective organisation. How can the song cycles be dated except by the year of translation into English? Very few people read an anthology from beginning to end, and those who do will find that in this one the alphabetical arrangement has thrown up some fascinating conjunctions. Anthologies can be approached in many ways. By preparing an index of themes as well as indices of first lines, and poets and titles, I hope I have aided people who like to use poetry in their liturgies or who find literature of help in their private meditations.

I should also say a word or two about how this book differs from formative and normative anthologies of Australian poetry. Jacques Derrida has a lovely image of how words seem when placed in quotation marks: 'like a garment spread out on a clothesline with clothespins'.[9] Thematic anthologies, like this one, do much the same thing; they take a familiar body of writing and place it in an odd configuration. I find it strange to have edited an anthology of

Australian poetry and not to have included anything by Kenneth Slessor who has surely written several of our finest lyrics and very likely our greatest long poem in 'Five Bells'. But I could not find any trace of religious feeling or thought in his writing, much as I admire it. With those poets well-known to Australian readers, there is sometimes little overlap between what other anthologists have chosen and what appears here. I hope that no one will conclude from this that Australian religious poems are not amongst our most memorable. The point is, rather, that a thematic look at a literature can uncover material that has been bypassed or overlooked.

Finally, a number of people helped me bring this anthology to completion. Without Peter Rose, who proposed the idea of the anthology to me, it would not exist. I am indebted to my research assistant, Bridget Bainbridge, who helped locate the several thousand volumes I needed to read or reread in order to make this selection. Patricia Excell helped me establish the correct text of Francis Webb's poems, and Margaret Clunies Ross offered expert advice on the songs of the Anbarra People of Arnhem Land. Djelal Kadir allowed me to explore some of my thoughts on Australian poetry in the journal he edits, *World Literature Today*, and traces of that piece may be found here. I am grateful to Robert Pargetter, former Dean of Monash University's Faculty of Arts, for making funds available for research assistance. I am obliged to Carol Freeman who photocopied my selections week in and week out, and to Katherine Steward for meticulous copy-editing. Robert Gray, Noel Rowe, Vivian Smith and Chris Wallace-Crabbe each read a draft of my contents pages and made useful suggestions. Paul Kane commented on an earlier version of this introduction. And last of all my wife, Rita, supported me throughout the days and nights of the whole project. My thanks to all.

Kevin Hart
Melbourne 1994

Notes

1 Limitation of space has prevented extensive citing of poems in their original languages. However, I have given Dimitris Tsaloumas's poem in Greek as well as English and the Anbarra People's chants in their native tongue.
2 Maurice Blanchot, *The Space of Literature*, trans. and intro. Ann Smock, University of Nebraska Press, Lincoln, 1982, 198.
3 Friedrich Nietzsche, *The Gay Science*, trans. Walter Kaufmann, Vintage Books, New York, 1974, 181.

4 Edmond Jabès, *The Book of Questions, I: The Book of Questions*, trans. Rosmarie Waldrop, Wesleyan University Press, Middletown, Conn., 1976, 138; *The Book of Questions: Yaël, Elya, Aely*, trans. Rosemary Waldrop, Wesleyan University Press, Middletown, Conn., 1983, 160.
5 Paul Auster, 'Book of the Dead: An Interview with Edmond Jabès', in *The Sin of the Book: Edmund Jabès*, ed. Eric Gould, University of Nebraska Press, Lincoln, 1985, 19.
6 Samuel Johnson, *Lives of the English Poets*, ed. G. B. Hill, Clarendon Press, Oxford, 1905, vol. 1, 291–2.
7 For full details, see Ian Gillman's *Many Faiths One Nation: A Guide to the Major Faiths and Denominations in Australia*, William Collins, Sydney, 1988.
8 Ludwig Wittgenstein, *Philosophical Investigations*, trans. G. E. M. Anscombe, Basil Blackwell, Oxford, 1972, §67.
9 Jacques Derrida, 'Living On: Border Lines', in *Deconstruction and Criticism*, ed. Geoffrey Hartman, Routledge & Kegan Paul, London, 1979, 76.

Robert Adamson

The River

A step taken, and all the world's before me.
Night so clear,

stars hang in the low branches,
small-fires, riding waves of thin atmosphere,

islands parting tide as meteors burn air.

Oysters powder to chalk in my hands.

A flying fox collides against my trunk
as the first memory unfurls.

Rocks on the shoreline milling the star-fire,
and each extinguished star,

an angel set free from the tide's long drive.

The memory shines—its fragments falling into place,
and the heavens revealing themselves

as my roots trail, deep nets
between channel and shoal, gathering in

cosmic spinel, Milky Way, Gemini.

I look all about, God, I search all around me.

She surrounds me here
as the light transfigures light

the butterfly exploding herself, colours thrown over
the nets of light in transfiguration.

The sea's adrift, tails outspread, the harbour dawn.
A gale in my hair as mountains move in.

I drift over lake, through surf-break
and valley. Shifting before me

another place. *On the edge or place inverted*
from Ocean starts another place,

before me—entangled of trees, unseemly
in this time, this place.

Humming nerves of the tide, the eels
twine themselves round, loop and flick

glow through valleys of silt, rise breaking surface,
twisting light, dislodging

memory from its original lineaments.

Tonight time is its own universe, shining in mangroves
through opaque leaves, bodies, plumage

and hands across tide.
The oldest fear returns through the monument of a fishbone,

and wings of an ice bird waving from rockface
with hardly an instinct.

A step back and my love's before me,
life shot through with these savage changes.

The memory ash—we face each other alone now
with no God to answer to.

After centuries, almost together now
the threshold in sight.

We turn in the rushing tide again and again to each other,
making fire of this, and setting it

here between swamp-flower and star.

To let love go forth to the world's end,
to set our lives in the Centre.

Though the tide turns the river back on itself,
and at its mouth, Ocean.

Anbarra People of Arnhem Land

Ngalalak

Wang-gurnga guya, wang-gurnga guya, gulob'arraja,
ngwar-ngwar larrya, maningala rarey Ngaljipa.
Jamburr bujarinya, blayriber larrya, garrarra-garrarra,
Ngwar-ngwar larrya, blayriber larrya, jamburr bujarinya,
Ngaljipa guya, garambak mbana.
Yeliliba guya, ngwar-ngwar larrya, garrarra-garrarra,
rarrchnga guya, blayriber larrya.
Ga-garrarra rarrchnga guya.
Ngaljipa guya, ngwar-ngwar worrya, jamburr bujarinya,
blayriber larrya, ngwar-ngwar worrya, maningala rarey,
rarrchnga guya, Gulgulnga guya.
Ngwar-ngwar worrya, yirpelaynbelayn, rarrchnga guya
Ngaljipa guya.
Ngwayrk, ngwayrk, Gulgulngam.

White Cockatoo

White cockatoo, white cockatoo gorging on grass seeds,
dancing and leaping in the sky
at Ngaljipa.
At his upland forest home he eats corms and dry grass
seeds,
his crest bobbing up and down.
He dances and leaps, greedy for grass seeds at his forest
home, at Ngaljipa where he plays didjeridu.
See him leap, his crest rising and falling, see him eat
rarrcha grass and corms!
See his crest bob up and down as he eats the *rarrcha* grass!
At Ngaljipa he dances, his crest bobbing, gorges and
belches, dances again and leaps, eats *rarrcha*
at his birthplace Gulgulnga;
See him dance in the sky, see him eat *rarrcha* grass at
Ngaljipa!
He calls *ngwayrk ngwayrk* at Gulgulnga.

PERFORMED BY FRANK MALKORDA, 1978
TRANSLATED BY MARGARET CLUNIES ROSS

Wama-Dupun

'Nga miwarl mbena, gungupa yoryordeya, nakurra wargolgolya,
barlbag-barlbag merney, biyorda ngurrmuryala, Garlnga miwarl yana.
Wurrjalab' lambirney, kolupanda yayey, Badurra yibirri-ibirr.
a la la la la
Jarpa murrayala.
Gungupa yordey-yorda, damerra wurrjalaba, a barlbag-barlbag merney, damerra wurrjalaba, kolupanda yibirri yibirr, Badurra yibirriyibam.

Wild Honey and Hollow Tree

Spirit women belonging to wild honey hang up their baskets at Garlnga, full of fat sugar bag, cut, waxy cells oozing dark, viscous honey.
Hollow tree, Wurrjalaba, Badurra hollow log coffin, dry wood,
Full of fat sugar bag, gathered by spirit women at Garlnga, dark honey—
a la la la la
Wild honey seeps and stains the dry tree trunk—
Fat sugar bag, oozing with dark, viscous honey, hot stuff,
oozing from dry wood, from Badurra hollow log coffin.

Performed by Frank Gurrmanamana, 1975
Translated by Margaret Clunies Ross

Muralkarra

Daunyiley-nyiley, gaya barrnga, gulbi birrirra warralanga, wardupalma, birrirra borja, Wakwakwak, jirnbangaya.
Birrirra borja, garma borja, Garanyula-nyula, Warduba jirnbanga.
Birrirra borja, wandalanga, gurta birrirala.
wak wak wak
Bianga borja, jirnbanga.
Badurra borja, wandalanga, a Maraychnga, daunyiley-nyiley,
Badurra Wardupalmam.

Crow

Crow plays and sings, rubs his firesticks together—see his
track in the skies!—gets up to dance and tap
his sticks—Wakwak's a dancing man.
Crow taps his sticks, perches on hollow log, dances at
Garanyula, his camp in the upland forest—
Wardupalma's his clan.
He climbs on Badurra—see his heavenly track! he's
dancing up above.
wak wak wak
A flock of crows caw to each other as they eat, then rise
to dance.
Crow perches on Badurra—see his heavenly track!—on
Maraych; he plays and sings, Wardupalma clansman
dancing on Badurra.

PERFORMED BY FRANK MALKORDA, 1982
TRANSLATED BY MARGARET CLUNIES ROSS

Dorothy Auchterlonie

Resurrection

And in that morning on the grey and sullen plain,
They heard the last notes of the trumpet wake the day,
And stretched their bones and shook the dust away,
Flexed their stiff shanks and stood erect again.

Their sockets danced with pain in the white light,
The new winds lashed each frail anatomy,
Roared in their ear-holes like an angry sea,
Shattered the peace of their millenial night,

Died down, and then in fury stirred afresh:
Each grinning skull grew fixed with sudden fear,
Felt the old agony of breath draw near,
The nameless terror of returning flesh.

They stretched their bony hands in silent dread
And wordless prayed the blank, unpitying sky . . .
But blood returned, with brain and tongue and eye
And space resounded with their bitter cry:
'Lord God have mercy, let the dead stay dead!'

‘William Baylebridge’

(William Blocksidge)

From *Life’s Testament*

II

God, the Unguessed, for clay to stay me,
  Rifled through remotest spheres;
God, to endow the breath that bade me
Rise, to nerve this heart, arrayed me
  From the untold in His.

Admitted to that Fact profound,
  Reverseless, I must travail till
I force me to my farthest bound—
Till universal clay be crowned,
  All godhead, by the will.

Through tense mutations I must fight
  To larger labours; fallen, I must
Uplift me with the god in sight;
And shout defiance when they smite
  Me, bleeding, to the dust.

E’er growing to the Greater Me
  In toil thus, I will shed despair,
Yea, all this mortal shed, and be,
Through battle borne to battle, free
  To hail the Unguessed One there.

III

Learn that thou art immortal—
And that the immortal,
That waits, nor could, upon death,
Is the sheer now.
Thus shalt thou come to generous blood,
Magnanimity stronger
Than the pride of kings,
Deeds that laugh sublimely
In the face of God.
Thus shalt thou rise to thought that takes
The still wider circle,
To thought that builds
Another outpost in the Universe.

'Salvation'

They knew, who had the sun and heavens in sight,
Only the agues of their inward night.
The priests came—and Credulity was roused
To think all space and time its plight espoused!

Deity

This, of divinity, believe—
No concept can its other-world achieve.
That godhead in the man we shadow limned
The unguessed, the abysmal, deity hath dimmed.
How shalt thou Man, his truth completed, call?
God, by his proxy on this destined ball:
God, written tersely in this animal.

Bruce Beaver

From *Lauds and Plaints*

III

To the memory of him who wrote the word ETERNITY *so carefully on Sydney's pavements.*

eternity on street corners
so obviously belonging in such
a setting

in a clerk's copperplated script
with surveyor's yellow crayon   the one
big word

nobody ever saw you at it
until near the end you were lumbered and warned
indulgently

to keep out of the way of young cops
who may not understand why an
elderly gent

in a grey dust coat would practise calligraphy
at just such a time in this suspicious
manner

none of our business whether you were that
paradox in a higher tax bracket
a single man

or if you had a home in some concrete
and grass suburb away from the black-board
bitumen

that took your message to heart   that vanished
with the tread of vanishing feet beneath
the juggernaut

time   thirty years ago
when too many young men and women
too soon

were finding out about your word
I first saw it on corner after
corner

of the grey by day blacked-out city
  when I pushed a messenger's trolley
  or lugged

an overload of parcels from warehouse
  to shop   fourteen years old   finished
  with schooling

in words   what has become known as
  a drop-out   but a poet already
  in intent

a home-made magician's kit of other men's
  spells sizzling and bubbling in
  my serpentine

mind   my tongue dumb for three more
  years until two place-names fierily
  atomised

in far-off Nippon fructified
  my early loving/hating liaison
  with words

yellow days of years   the one
  real word in an unreal world
  eternity

I knew it was here already surrounding
  us   adding to subtracting from
  our moments

crossing and dotting our *I*s and lives
  with its big beautiful script looping
  volute

that was before I learned of the commonest
  agony of all   time's rape of
  the timeless

as I watched the generations of dogs
  excreting religiously over it
  the myriad

leather soles taking a little of it
  with them into homes shops
  offices

and the closest thing to sanctuaries
  of grass stone leaf sand
  wave-lapped

rock   out of the streets and into
  their lives blindly underfoot
  always

say that you had retired from packing-bench
  shop-counter   something modest
  enough

inflating skimped dreams with such
  grandeur of forever   so conscious
  a glimpse

of that *other* we hover about as a gnat-
  cloud of scintillant lives in a shaft
  of sun

if you meant an inkling of hell that's not
  your fault   familiarity
  breeds fear

as often as contempt   pain's ever-
  present to help us stay awake
  the big sleep

of the partly living isn't quite death
  for wherever life is is
  forgetfulness

and memory   remember yourselves
  you said over and over for over
  thirty

livelong years   not long in eternity
  yet what statistics of shoe-leather
  knee-bending

miles of yellow crayon stretching
  from Sydney to Parramatta to
  eternity

towards the close of your circuit you came
  into the salty streets of Manly
  the lettering

fainter   less forceful   a little shaky
  but still as clear as ever the word
  forever

did you pay the ferry man with those four
  syllables and did he splash
  the while

impatiently with an oar   a dream
  on a boat trip back to the darker side
  of the harbour

on this much lauded isthmus between
  the tides that rose and lapsed before
  your word

was ever enunciated the seasons
  stir and flex   go rigid or lurch
  swashbuckling

through the perennial montage   now
  is forever three years after you
  wrote finis

to infinity   but there's always a message
  in all our frenetic or soporific
  actions

even in this my noising of
  these all-too-soluble somethings of utterance
  into

the silence you now inhabit   just
  to say haltingly somehow or other
  the message

inescapably got there   got over
  to a number for what it was pricelessly worth
  your chapter

of obsessional effort   our paragraphs of
  approaching and passing   the one real word
  remaining

Judith Beveridge

The Herons

Then the path wound down
to a browner place, to a river
where rain-grey herons slender as rushes
drifted off like camp-smoke.

I've only seen their colour
in a few opals baked deep in clay country.
When they stared, it was as if
their eyes carried on

through emanations.
One stood so peacefully
as if it saw and heard the single
far off, crystal note;

slender, rag-thin bird we called
blue Gotama. We crumbled a mushroom—
all we could call
sacred, yet common:

but they looked past all hungers.
So we trod quietly back,
left them sitting above the long
brown earthworm of the river

and our pile of useless
vegetable soil. They were
beautiful as blue veins in the wrists of monks
fasting for perfection.

The Dispossessed Angels

Now, we are grounded. Once we were
like windmills turning on fertile slopes.
The air made us dizzy, spun our heads like propellers.
Weathervanes lifted us off our feet.
We were the powersource that saved souls!

Now, we don't even know where the water is,
or what to do with the wind, except
hoist it high in our cheeks and keep it there.
People, they say your mountains breathe
like bedridden invalids. Is it true—

will we go up to those hushed peaks?
We have heard how in the high mountains
breath is white light marbled with subtle oils.
And isn't breath something to stir the silver
alive, a wish for an angel?

In a ravine now, we sit remembering—
how we lost our foothold and plummeted.
Still our hearts uncork in our ears—
as if at some terrifying level. And we don't
breathe easily. We pant looking up
at those terrible and trackless slopes.

Blue vistas may revive us and peaks
gently resuscitate true breathing.
But we wonder by what fragile cohesion
the body keeps the soul floating
above a treacherous landscape.

Often we've watched ourselves: as if we were
tiny bubbles levitating over a hair brush.
So what do we do, trapped in the ground state
of our terror? In the soul's slipstream
maybe we'll lift a few inches.

But now we follow the flat drawl of the horizon
(though our power weakens the closer to sea-level).
O, to look down or up and not panic.
Do you know how it is? Then imagine how it is:
our glorioles blackened by fossil fuel
when they could have been brilliant and blue.

Performing Angels

When the vestibule's all light
& stars sway like a night of sung carols—
can you hear us?

our clear castrati voices as we bump
into the props
of old morality plays,
the seraphim cooing
for the soft focus of cathedrals.

We try all night
to get the sounds right,
God singing directions
like an overzealous catechist,
the cherubim as if squeezed through a piccolo.

Those old akashic tapes:
voice upon voice upon voice upon voice:
a choir of ages.
Sometimes you'll hear
the operatics of a requiem

through your thicker air,
here attenuated to Mozart, lighter
breakfast music. (Short of decibels
the idea of rock's exotic—but unpentecostal
as a gravelly-voiced angel).

Hour after hour we practise
the exact timbre
of a drum roll that in our thinner atmosphere
goes soft, quiet
as time-pips.
Our place on the eternal scale
is as measurement—

note by pinnacle note
against the seismic-sound of the occult,
or psyches tuned so low off the known scale
& trembling with a terrible stage-fright
under the spot-lights of our halos.

We sing!
just one cycle per lifetime
but our range is infinite.
Hear us penny-whistle
the hour: ourselves
and the lovely Rehearsing Ages.

John Blight

It

It is outside and cannot come
into this mind without these words;
though listening be wearisome,
it cannot enter without words.
Gently at first—it cannot burst
the scarlet shutters of this brain.
The eardrums it must rumble first;
stand at the eyes' blue windowpane.
And when at length it sees a light
shining within the intellect
it then may stand, as shadow might,
outside the entrance, tall, erect;
awaiting there in dread and fear
that she within may turn no key
unwilling yet to see or hear
unravelling of this mystery . . .
Then, meaningless, may traipse and pass
across the garden plot, the grass;
and, standing at the gate, may be
for ever afterwards a tree.

Ant, Fish and Angel

Part of me, in the morning, may be an ant
or a fish swimming away, as I spit,
or defecate, collectively fouling the bay.
So I am part of my world and can't
escape the bare truth, I am part of it;
that, somewhere, an ant, or a fish swimming away,
is a part of me. Oh, ant! oh, circling fish!
stay, and look hard at me. Is it your wish
to be part of man, to devour his innate fear?
Into the maw of an ant I disappear.
How trifling—be it a minnow's appetite
or some great fish in a more sizeable bite
disposes of me—as such to reappear!
Angels and gods come bite and take your share.

Deities

I could conceive of a God who
loved perfection; but never of a
monster-warrior in the skies
warring upon mortal mores,
to bend to this will or to that.

A God I'd talk with would live on the
Earth, as I ever amazed at
beauties of far constellations,
their likenesses in the countless suns
that breach upon a wave inshore.

I'd differ from him, when I chose
a colour and, through diffusion, we
would paint again the perfect smile.
The wonder, never mine alone,
this God would be the artist, not

the cancerous scientist who
tugs the Universe apart. No
need for deeper scrutiny than
a child's! That would be his first and
last love of creation: to

touch, to smell, to taste, to hear and
see; quite undisturbed by nightmare,
living in the day, loving each
tumbling second beautiful in
the waterfall—last, to look up
to see the stream was Life.

Francis Brabazon

Well have you called yourself the Ocean of Mercy—

Well have you called yourself the Ocean of Mercy—
For your shadow the sea has now rejected us,

Flinging us up on this inhospitable beach
Without even the ragged sail which protected us.

We know that a thousand times we have disobeyed you,
And a thousand times you have lovingly corrected us.

But it was not that we wilfully turned aside,
But a sickle shape promising reaping that deflected us.

Our greatest error, Beloved, was our presumption
That out of this teeming world you had selected us

To carry your message and sing your songs in the sun,
And our secret desire that men respected us.

How faithful to you is your shadow, even this sea;
Impartially it has judged and rejected us.

Christopher Brennan

Farewell, the pleasant harbourage of Faith

Farewell, the pleasant harbourage of Faith
and calm repose 'neath sunny skies and bright—
or was it darkness vainly thought the light
and all we worship'd but a fleeting wraith?
Me from that haven with vexation fraught
Doubt drives to wander: in adventurous bark
I follow e'er 'neath louring skies and dark
o'er gulfs of gloom and misty seas of Thought
upon their oar-blades' vanish't track who sped
to greet the rising sun if sun there be—
yet never unto them that light was shown
nor ever since these mingl'd with the dead
hath sun arisen on that shoreless sea
or man won way into the vast Unknown.

From *Towards the Source*

VII

The grand cortège of glory and youth is gone
flaunt standards, and the flood of brazen tone:
I alone linger, a regretful guest,
here where the hostelry has crumbled down,
emptied of warmth and life, and the little town
lies cold and ruin'd, all its bravery done,
wind-blown, wind-blown, where not even dust may rest.
No cymbal-clash warms the chill air: the way
lies stretch'd beneath a slanting afternoon,
the which no piled pyres of the slaughter'd sun,
no silver sheen of eve shall follow: Day,
ta'en at the throat and choked, in the huge slum
o' the common world, shall fall across the coast,
yellow and bloodless, not a wound to boast.
But if this bare-blown waste refuse me home
and if the skies wither my vesper-flight,
'twere well to creep, or ever livid night
wrap the disquiet earth in horror, back
where the old church stands on our morning's track,

and in the iron-entrellis'd choir, among
rust tombs and blazons, where an isle of light
is bosom'd in the friendly gloom, devise
proud anthems in a long forgotten tongue:
so cozening youth's despair o'er joy that dies.

Vincent Buckley

Before Pentecost

My soul has learned the country of her fear
Yet doubts, and lingers there
In the prison of her disorder, where a point
Of pain finds every joint
And the nerves live only for grievances.
I live where hell is,
Peaceful only till the chill drops start
That sting my limbs and heart,
Or the great veins throughout my body beat
And the hands crawling with heat
Grow transfixed by their own nails. Then where
Shall I find strength to bear
This burning shade, this cramped and cruel retreat?

Under the darkening pressure of the sky
All things consent and die.
And there is something drives me, more and more,
To whisper and explore
With my nerve-ends the continent of night
Till, far from the Blessed Sight,
Even that withers. Hell is only a name
For this devouring shame
Which holds me focused as love never could,
Blight of the green wood,
My friends the friendless emissaries of the dead,
The living earth my dread
And worst of goads, this freezing sweat my blood.

Is it possible I have served my Lord so ill?
Or that the mounting will
Should fall so far and sudden? There's no trace
Of summer now, no face
That brightens in this landscape or in my soul.
Christ, from pole to pole
The earth lies rich and level as a dream;
And I, I cannot seem
More than a phantom to myself. All choice
Lies frozen at the source.
Must I go dry and numb another year
Before my senses bear
His footstep, and the kind accusing Voice?

Song For Resurrection Day

O world I covet, earth of heaven,
How shall your fields compel my sight?
In whom love is a fiery oven,
And sleep a torment of the night.

And yet I know these bones will move,
These eyes will see the heavens walking.
Then, heart, grow molten with your love;
Cleave, soul and body, to that waking.

Puritan Poet Reel

Mother at her novenas
Until her knees are brown
And father non-conforming
All around the town:

What images have brought
(Ah delicate and dry)
The tear of ambiguity
To stand within my eye?

Have built my world around me
Where, private as a mole,
I guard the fiercer virtues
And mentally control

Wind squeezing the houses,
Knocking the hedges down,
And father non-conforming
All around the town?

Mother it was that promised me
Position in the town
And father raised his fist and swore
We'd lie on beds of down:

But who will keep the promises
They made me as a boy?
Writing my twenty lines a day,
And simulating joy,

I make my life a model
And keep my bowels clear,
But, muse blow hot or muse blow cold,
Over the fence I hear

Mother at her rosary
Until her knees are brown
And father non-conforming
All around the town.

Places

(For J. Golden, S.J.)

I

Walking at ease where the great houses
Shelter the assorted trees that someone
Planted, once, to shelter them, I do
My voluntary patrol. The wind moves
Houses and trees together, till they breathe
As though I breathed with them, systole, diastole
Of the built and the growing.
Fair enough. We used to picture
Paradise both as a garden and a city.
Here it's a green hardihood, a tender
Rallying beyond concupiscence.
So I patrol. There's not a soul in sight.
It was an older, foreign voice that cried
'The swarm of bees enfolds the ancient hive'.

II

But love is a harsh and pure honey.
The world is brought alive with us
So many times. One night I learned the resurrection
In still water. Sea-mist moves
On a land that in its steeped
Peach-dark fruits,
Resin,
Pods,
Is warm as blood.

I lean on the bridge, looking down.
Under the utter moon all things reach
Their height in water; there the thin
Unbreathing tree touches the depth of cloud
Downward; there light vibrates in the sky.
In this voluptuous arrest of colour
I still feel the day's heat on my eyelids.
At noon the summer webbed us in; but now
I almost smell the next year's seed.

III

Bound from Mass, my blood fresh as the sea.
In the city light there are pools, deep-groined,
Where the gilled bodies leap down and glide;
And the sea-smell, drifting like the sounds of sleep,
Gives air a distance, not a shape,
And light itself is recreated, made
Native to all bodies. I think how once,
Hardly thinking, in a strange church,
A man, forgetting the common rubric, prayed
'O God, make me worthy of the world',
And felt his own silence sting his tongue.

From *Eleven Political Poems*

No new thing

No new thing under the sun:
The virtuous who prefer the dark;
Fools knighted; the brave undone;
The athletes at their killing work;
The tender-hearts who step in blood;
The sensitive paralysed in a mood;
The clerks who rubber-stamp our deaths,
Executors of death's estate;
Poets who count their dying breaths;
Lovers who pledge undying hate;
The self-made and self-ruined men;
The envious with the strength of ten.

They crowd in nightmares to my side,
Enlisting even private pain
In some world-plan of suicide:
Man, gutted and obedient man,
Who turns his coat when he is told,
Faithless to our shining world.
And hard-faced men, who beat the drum
To call me to this Cause or that,
Those heirs of someone else's tomb,
Can't see the sweeter work I'm at,
The building of the honeycomb.

Day with its dry persistence

In day with its dry persistence
In night warm with the first star
Down the midnight-passages
Or in the small corners of silence
Or at the bedside hot with death
A restlessness that clings and will not
Be rubbed off on paper.

Yet there are some tempos that prefer me,
Some twigs that burst with shaking
Blossom and dew, some lights that are constant,
Some movements of the earth that bring me
In constant pilgrimage to Genesis,
To the bright shapes and the true names,

Oh my Lord.

Ada Cambridge

Vows

Nay, ask me not. I would not dare pretend
 To constant passion and a life-long trust.
 They will desert thee, if indeed they must.
How can we guess what Destiny will send—
Smiles of fair fortune, or black storms to rend
 What even now is shaken by a gust?
 The fire will burn, or it will die in dust.
We cannot tell until the final end.

And never vow was forged that could confine
 Aught but the body of the thing whereon
Its pledge was stamped. The inner soul divine,
 That thinks of going, is already gone.
When faith and love need bolts upon the door,
Faith is not faith, and love abides no more.

David Campbell

Far Other Worlds

A white wind turns the daylight moon
And Time is whittled on that stone
Where fat heart sits when he is fed
And whets his grief when he is lean.

The living light upon that ice
Is transient amber in saints' eyes;
There lovers wonder and, love gone,
Grief's torture-pointed comets rise.

There flares the tree the mystics burn
And there is bent the tree of thorn
And griefs dog-buried in that land
Shall stand up green between the stone.

Speak with the Sun

From a wreck of tree in the wash of night
Glory, glory, sings the bird;
Across ten thousand years of light
His creative voice is heard.

Wide on a tide of wind are set
Warp and woof of silvered air;
But the song slips through the net
To where the myriad galaxies are.

And to the heartbeats of the light,
Now from the deepness of the glade
Well up the bubbles of delight:
Of such stuff the stars were made.

The Miracle of Mullion Hill

(To Jock Maxwell)

The cock has made his winter perch
The roof-tree of the iron church
And straining heavenward on his toes,
Turns scarlet, mops his wings and crows;

At which the Reverend Father Pat
Rolled out of bed, but reasoning that
The Lord who fashioned flesh and frost
And died that man might not be lost,
Would surely not expect his heirs
To catch their death while saying prayers,
He tumbled back to bed again
And prayed, 'Dear God, first send us rain;
And then, should you see your way clear,
Rid me of the cross I bear!
Relieve me, dear Lord, of Hanrahan,
The parish wood-and-water man,
Or else return him to the fold.'
And reaching out into the cold
To cross the blankets on the bed,
He heard the cock cry overhead.
Then in his lean-to at the back
Hanrahan woke, and to a crack
Applied an eye, a shepherd's warning,
And seeing his reverence at morning
Prayer, he banged about the room
Carolling cheerily 'Rolling Home'—
A ballad which, to say the least,
Is not fit matins for a priest.

'O Hanrahan! O Hanrahan!
Fear for your soul! I tell you, man,
This will not do! What, laying odds
And drinking late? Can these be God's
Recipes for heaven? Well,
Go to the . . . go and ring the bell
Calling the pious and the just
To Mass. And may the Holy Ghost
Descend upon you, armed with light,
And blast a pathway through your night
That shows more bleak and stubborn than
This winter morning, Hanrahan.'

So saying, Pat returned to prayer;
And on the crisp and holy air
The bells filed out like sheep along
A mountain pad, ding dong, ding dong.
Ding-dong, ding-dong! In peals and volleys
They skipped like rams across the valleys,
To saunter home in single file
Leading the faithful up the aisle.

Now Hanrahan, you may have guessed,
Was one who liked his little jest
But did not care for Father Pat.
He'd taken up a contract that
He, Hanrahan, the undersigned,
Would clear such lands as were confined
By certain parallels and degrees,
Felling and stacking forest-trees
Upon that portion of the chart
Named Mullion Hill or any part
Thereof adjacent to the kirk;
With paragraphs on week-end work;—
A ruse devised by Paddy Ryven
To settle certain scores with Heaven.

So while the eucharistic bell
Tinkled like frost on Mullion Hill
And *Glory, glory, God on High!*
Rose to the green and luminous sky
On tongue of bird and tongue of man,
This sabbath-breaker, Hanrahan,
Shouldered his lean short-handled axe
Stained with the blood of chicken-necks,
And set to work. Quite soon the air
Was thick with chips; and if sweet prayer
And cries of 'Timber!' do not mix,
What's that to him? Why less than nix—
For were not his occasions lawful?
But Hanrahan, you'd best be careful!

For, as he paused to flick the sweat
Out of his eyes and spread his feet,
He heard high up on Mullion Hill
The chiming of another bell,
A bell that trembled in the hush
Like winter water through the bush;
And then, as if the mountain spoke,
From orchid-tongues and granite rock
There burst a litany of praise
And alleluias! 'Spare my days!'
Cried Hanrahan. 'Another Mass!'
And stumbling through bracken, stone, and moss,
He followed the vanishing service up
Gully and cliff until the top
Of Mullion rose against the dawn—
And still the tinkling bell led on.

This was not strange, for he had heard
The mimicry of the lyrebird
Who kept a mound close by the fence
And stole the music for his dance
When tail a-tremble through the fern
He chased his bright-eyed lyre-hen.

It chanced that high on Mullion Hill
A ram was grazing. Two years' wool
Hung from his brisket like a beard;
He had a grave, a noble head;
And from a precipice he looked down
Majestically. The early sun
Leaping that moment from the east,
Gilded the shoulders of the beast
And set with jewels his horned crown
And so he gazed at Hanrahan—
Who falling down upon his face,
Cried, 'Heavenly Father, send me grace!
Pardon my sins, Lord God of Hosts;
I never stole the Holy Ghost's
Communion wine; and as for trees,
I'll cut that out. Here, on my knees,
Behold your humble servant, God.'
At which the vision seemed to nod.

So now when star-frosts glitter round
And bells ring out and hills resound
Calling to sinners to repent
And join with them in sacrament,
Who is it hands around the plate
And frowns should anyone be late
And swings the censer up and down?
Who but holy Hanrahan?

Fisherman's Song

There I would cast my fly
Where the swan banks and follows,
Though stars are foxed, the dry,
The vanished river's shallows—
And all of time in her cry.

By rock and silted bend
Where the buried river ran
And grass sings in the wind,
I would follow the swan
To the reach of her mind—

Till rock and mirage break
And stars double and float
Upon the quiet lake.
There I'd put out my boat
As the herons wake,

And tossing to the floor
An empty spindle,
I'd rest upon an oar
Watching the dawnlight kindle
Christ's fire on the lake shore.

Among the Farms

May He who sent His only Son
To torment on a cross of wood,
Look twice on animal and man
Caught in the narrow ways of blood.
The moonlit fox who hunts by night
The lambs that dance these frosted hills,
Hunts first from need and appetite,
Then for his own delight he kills.
I knew a man with violet eyes,
A countryman who loved his ewes
And swore, by God, he'd put him wise,
And when the fox was in his noose,
He stripped the russet pelt for prize
And set the living creature loose.

From *Cocky's Calendar*

Prayer for Rain

Sweet rain, bless our windy farm,
Stepping round in skirts of storm
While these marble acres lie
Open to an empty sky.

Sown deep, the oaten grain
Awaits, as words wait in the brain,
Your release that out of dew
It may make the world anew.

Sweet rain, bless our windy farm,
Stepping round in skirts of storm:
Amongst the broken clods the hare
Folds his ears like hands in prayer.

Trawlers

Sun orchids and wild iris as violet-blue
As the wine-dark sea; and a death below, the ocean
Unfolds silk petals on rock—bolts thrown on
Aphrodite's table. But death is cloaked, is true,
On this eminence. Close sleepless eyes and you
Are in the old nightmare. In imagination
Gloucester stood here beside Poor Tom, his son,
And threw himself down to find, as most men do,
That he must put up with blindness and dishonour.
Rounding the lighthouse come the trawlers sailing
To their rusty haven. In baskets, flowers of the sea
Are stacked in quicksilver tiers for gutting, scaling;
And the John Dory at this silver hour
Still bear Christ's thumb-marks, his blessing at Galilee.

A Yellow Rose

From the churn
Yellow butter
Arrested light
Beaten into a rose

Light unfolds
Slowly
A savage love
I burn I overflow
The bee's bite is endless

The rose moves with time
A yellow beak
My blond love
A floral madam
Shaken by laughter

Mass
At the speed of light
Is a yellow wafer

To the stationary observer

Timeless petals
Blow on the wind

Gary Catalano

The Jews Speak in Heaven

We stood at the edge
of the gouged pit

and looked down
into history. With all

those unbroken clods
at its edge

it looked like
a tin-can, hastily

torn open.
As we listened

to new metal
slipping into

the greased bolts
how could we

reproach the sky
when it showed us

no Jehovah? There
—there was the warmth

of known
stretchmarked flesh

and there, too,
the strains of a

familiar tune
ringing

in our heads. Soon it became
so loud

our ears
even bled.

A Dream of Hell

Fire? Or endless mud? No, I see neither of these? I see a landscape of torn and broken books. Here you can crouch for a thousand years and piece together the pages of the one book which bears your name, and thereby learn the meaning of your life, yet here it is too late to take advantage of that knowledge. Alive, why didn't you think of reaching out . . . and of running your light hand across the braille of the stars?

Heaven of Rags

(after Eugène Atget)

The camera, as always,
has caught him
exactly—right foot

extended
to a point
almost the future, eyes

peering out
from a time
always the past. Between

these two points
he lives a thin
eternal life

in this photograph,
a man with no name
and an occupation

hardly pursued. Is it
wholly beneath our dignity
to collect rags, and so

maintain a bond
with a world
always there

a world that records
all its history
in these

pathetic rags?
I should imagine
that the true heaven

is a heaven of rags:
all things aged
beaten and torn

will be there, stained
with sweat and dirt
and basking in the glory

that only comes
of use—old hats
old boots

and the old
crazed skin
of the fallen

a skin so at home
on its body
it cannot

be exchanged
in a heaven of silk,
a heaven of rags.

Equation

It is only when we are completely convinced of the obstinate otherness of things that the cucumbers will emerge from their hiding places and show us their unbandaged wounds. Equally, it is only when we agree to loosen our grip on those steel bars in our heads that the lighthouse will reveal itself to be a pillar of pure lightning.

Hole

The woman has a hole in her face, a hole whose exact location cannot be pin-pointed. But it is there all the same. One minute you will be staring at her tanned and corrugated face and, quite suddenly, the hole will appear.

It has never appeared in the same place twice. One afternoon I gazed at her from across a table and saw it down near the corner of her mouth; on another I sat in exactly the same place and found it almost in the centre of her forehead.

Those who flinch and turn away on seeing the hole do not know what they are missing. If they focus on it and refrain from blinking, they may see another face within the hole.

I have seen this other face and know it to be the one the woman wore when she was a young girl. But having said that, I suspect it is also the one she will wear when the hole expands to its fullest extent and, quite suddenly, she finds herself standing—or seated—before God.

Alison Clark

Credo

The lights of a ship on the horizon at three a.m.
seem so significant, at the low ebb of everything
and time of questions: where we're going to and coming from.

A container ship on its run between cities,
but how it speaks from the blank diameter of greys!
We must be firm, regard our *heart only as shish kebab;*

yet who but some crazy mystic could be so foolish
as to keep on changing—trusting the words to catch up
and (in a charged hour) make their oblique syntheses?

It's like being the guardian of a madman
and out of the darkness he smiles and says your name.
You think: I am chained, but share the soul

which lights up everything. Even though, often,
it's just a creature scratching at the woodwork
or calling: *Bip! Bip!* like a bandicoot in the night.

Ananke

This is the dwelling of Necessity,
this body of the lake spread out at night,
dark volume on its bleached ellipse—
estrangingly so perfect, and itself;
lights of a fisherman and distant ship
facts in the sepia scene, promises of nothing.

And so I name the goddess
of the realm of slow days ruled by thankless tasks;
or scenes between attracted
characters from different plays,
forced by their scripts
to act the bleak bonhomie of a farce—

other, yet herself in us: the strong
law of dreams; inverse perspective
within each event which frames our next step;
indelible as nasturtium red;
intractable as the distances of stars
(not in celebration) I acknowledge her!

Respecting the Mysteries

Walking at night beside the lake
you note the haunting face trees
and tracks wear in moonlight and mist—

as when, speaking alone with someone,
we might sense another presence
in the room: a veritable being, and not

(just) the high degree of our own attentiveness;
nor old friend Pathetic Fallacy
who paints the world with our features—

siren leading us astray.
Easier, though, to say not what it is
but is not in the room, in the night with us.

Gardening

Digging planting watching buds emerge,
but specially watering, how comforting it is.
Like one of those looked for dreams
which resolves anxieties:
I turn the tap and my heart lightens
with the cool spray in the cool air,
glistening leaves—quenching,
consoling . . . I don't understand it.

It's like dowsing yourself
in sandy rock pools at low tide:
tossed between sea and sky, object
in the world of objects' play
and with the gift to know it,
the parched soul feels requited
as the walker in the city on a glary day
who enters a narrow street of shade.

It's like pulling away from busyness
and news you don't really care about
(not yours, yet you suffer because it is)
to read in the attic where none can see
you weep with love and envy
of the true exactly right odd words—
as if in a chain of small oases
happening meaning feeling coincide.

David Curzon

From *Midrashim*

> Blessed is the man who walks not in the counsel of the wicked, nor stands in the way of sinners, nor sits in the seat of scoffers; 2 but his delight is in the law of the LORD, and on his law he meditates day and night. 3 He is like a tree planted by streams of water, that yields its fruit in its season, and its leaf does not wither. In all that he does, he prospers. 4 The wicked are not so, but are like chaff which the wind drives away. 5 Therefore the wicked will not stand in the judgement, nor sinners in the congregation of the righteous; 6 for the LORD knows the way of the righteous, but the way of the wicked will perish.
>
> TRANSLATION: REVISED STANDARD VERSION

Psalm 1

Blessed is the man not born
in Lodz in the wrong decade,
who walks not in tree-lined shade
like my father's father in this photo, *nor*
stands in the way of sinners waiting for
his yellow star,
nor sits, if he could sit, in their cattle car,

but his delight is being born
as I was, in Australia, far away,
and on God's law he meditates night and day.

He is like a tree that's granted
the land where it is planted,
that yields its fruit by reason
of sun and rain in season.

The wicked are not so, they
burn their uniforms and walk away.

Therefore the wicked are like Cain
who offered fruit which God chose to disdain.

And *the way of the righteous* is Abel's, whose
slaughtered lambs God chose to choose
and who was murdered anyway.

6 Go to the ant, thou sluggard; consider her ways, and be wise: 7 Which having no guide, overseer, or ruler, 8 Provideth her meat in the summer, and gathereth her food in the harvest.

TRANSLATION: KING JAMES VERSION

Proverbs 6:6

Go to the ant, you sluggard,

and watch it lug an object
forward single file
with no short breaks for
coffee, gossip, a croissant,

and no stopping to apostrophize
blossom, by-passed because
pollen is not its job,
no pause for trampled companions:

consider her ways—and be content.

Bruce Dawe

Happiness is the art of being broken

Happiness is the art of being broken
With least sound. The old, whom circumstance
Has ground smooth as green bottle-glass
On the sea's furious grindstone, very often
Practice it to perfection. (For them, death
Is the one definitive shrug
In an infinite series, all prior gestures
Take relevance from this, as much express
Sorrow for stiff canary or cold son.)

Always the first fragmentation
Stirs us to fear . . . Beyond that point
We learn where we belong, in what uncaring
Complex depths we roll, lashed by light,
Tumbling in anemone-dazzled fathoms
Seek innocence in surrender,
Senility an ironic art of charity
Easing the agony of disparateness until
That day when, all identity lost, we serve
As curios for children roaming beaches,
Makeshift monocles through which they view
The same green transitory world we also knew.

And a Good Friday Was Had by All

You men there, keep those women back
and God Almighty he laid down
on the crossed timber and old Silenus
my offsider looked at me as if to say
nice work for soldiers, your mind's not your own
once you sign that dotted line Ave Caesar
and all that malarkey Imperator Rex

well this Nazarene
didn't make it any easier
really—not like the ones
who kick up a fuss so you can
do your block and take it out on them
Silenus

held the spikes steady and I let fly
with the sledge-hammer, not looking
on the downswing trying hard not to hear
over the women's wailing the bones give way
the iron shocking the dumb wood.

Orders is orders, I said after it was over
nothing personal you understand—we had a
drill-sergeant once thought he was God but he wasn't
a patch on you

then we hauled on the ropes
and he rose in the hot air
like a diver just leaving the springboard, arms spread
so it seemed
over the whole damned creation
over the big men who must have had it in for him
and the curious ones who'll watch anything if it's free
with only the usual women caring anywhere
and a blind man in tears.

Bring out Your Christs

We tried to hold him back
but he was strong (they get
like that) broke clear
and headed at a run
into the blaze,
                    the flames
enfolded him in their orange
arms . . .
            the building fell, well, anyway,
those who were saved
were saved, the rest
blended like him
into the mother ash that clung
to our wet gum-boots as we stepped across
non-existent thresholds, like dream-walkers strolled
through walls and smelled
the black smell of extinction.
The bloody fool, I said.
What was he up to?
                              Here, use the hose, said Jack.
You wouldn't want to be carting him everywhere.

A Week's Grace

Towards morning I fall asleep
swiftly as ash, my map-of-Australia profile
raised to the reddening sun.
Later I pinch my spirit awake
with a quick look in the mirror where every speck
has its twin to forestall loneliness.
In the shower I beat my body with water,
feeling the skin sigh, recounting
a week that went grievously well.
On the last night
the rain said insistently: You and your dreams!
And I must have turned over, mumbling,
making vague gestures out of myself
in the hope that finally the heart
would find its way home, now fast, now slow,
but getting there, nosing the bracken, head down, swinging
this way and that, a coat full of burrs,
travelling, despairing, singing.

At Mass

(for Peter Phelan)

Now, you're more one of us than we are, splashing
down from the dark side of the moon whose nautilus sheen
images forth our Church, reflecting the State's
pre-eminence, nightly, fading on cue
like a faithful consort, now you suffer the crushing
pressure of withdrawal, your soutane
changed for a business-suit whose grey completes
the metamorphosis, having failed to toe
the Church's business-line, your stand a cry
from this southern Church of Silence where to speak
what's in the heart is some dishonour . . .
Days
reel us in . . .
Snap-frozen in some seminary,
the Word, secured against the ubiquitous shock
of honest air or breath, rots as it thaws.

The Christ of the Abyss

In the waters off Portofino
the Christ of the Abyss
holds wide His arms
to the coral-divers drowning with their last baskets of red coral
around their necks,
to the spear-fishermen's fish, the amber-jack twisting on the bright
barb,
to the tourists' polaroid lenses spinning down through the lime
waves like thrown coins
—like the Christ of the Upper Air, the Christ of the Andes,
He draws all men unto Him, the little souls
circling in like fish
with their cries and concerns
mute gleams in the wanlight . . .

And the waters go over Him like air,
and like the fragile ancient amphora
bringing wine and grain to the great sea's shores
He outlasts both timber and iron, trireme and motor-cruiser,
the salt,
and the weed,
and the slime.

Rosemary Dobson

Two Visions

How strange to lie upon a bed
With hands and feet disposed for death
And smooth-shut eyes, while overhead
A flight of angels fills the ear
With, stainless-sweet and water-clear,
Blown trumpets sounded by their breath.

'Saint Anthony, how still you lie!
They are not dreams,' the painter said,
'Those bright frequenters of the sky,
They live, they breathe, and I should know,'
(Dabbing his brush in indigo)
'I placed them there above your bed.'

Saint Anthony, that holy man,
Sleeps on, and with his inward eye
He sees, as but the holy can,
The heavenly host, the wings, the Dove,
The Hand in blessing raised above;
And puts the world of living by.

'Well, dream your dreams, Saint Anthony,
We have our visions, you and I,
And who can say where Truth may be—
In holy dreams, or in my paint
Which shows the wonder and the saint—
Since both can pierce beyond the sky.'

The Missal

The lovely, lost, italic curve
Enrapt his soul. All day he stared
At *A* for Apple. Dimly heard
The beating of the holy wings
In *D* for Dove; but shut his ears
To sound's delight, instead preferred
The *E* of Eden, marked how stood
The mainstroke like a flaming sword.
Then *F* for Fall. How brief the flight
With dark descender, black as night!

The circumscribed, enchanting sweep
Of *G* was Goodness. Calm and strong
Gently it led his eye along
Delicious pastures of the page.
Then, overleaf, before him burned
The winged and serifed letter *H*
And Heaven was his to have and hold—
Oh missal wrought in holy gold.

The arrogance of earthly power—
Swash capitals for Kings and Men—
Diminished as he turned again
And summoned all his art to raise
The lettered record of his praise—
'Ere there was beast or fish or bird,
Before the Voice of Man was heard;
In the beginning was the Word.'

Callers at the House

While the doctor and the district nurse joined battle
with the recalcitrant illness of my mother

I attended to the questions of the trainee assistant,
young, and with candid eyes as blue as water.

In an adjoining room we sat conversing:
he was to be either a doctor or a theologian

and waited a blinding flash from heaven as signal.
Meanwhile he was collecting the Sermons of Doctor Spurgeon.

In which edition, I asked. In the earliest,
for the spines are stronger and the binding better.

My mother is eighty-three. Her spine is
more upright, her spirit more stoic than her daughter's.

Thirty volumes of the Sermons of Doctor Spurgeon!
She would eagerly share a pleasure in this collection.

Clear-sighted Athene who in the midst of battle
dispels the mist from the eyes of those she favours

visited this house in a brilliant flash. I heard my
mother speak to the doctor. 'I feel better.'

Near Epidauros is the shrine that Diomedes
dedicated to Clear-sighted Athene. I will go there

and pay my debt of thankfulness. Or should I
offer my thanks to the God of Doctor Spurgeon?

Being Called For

Come in at the low-silled window,
Enter by the door through the vine-leaves
Growing over the lintel. I have hung bells at the
Window to be stirred by the breath of your
Coming, which may be at any season.

In winter the snow throws
Light on the ceiling. If you come in winter
There will be a blue shadow before you
Cast on the threshold.

In summer an eddying of white dust
And a brightness falling between the leaves.

When you come I am ready: only, uncertain—
Shall we be leaving at once on another journey?
I would like first to write it all down and leave the pages
On the table weighted with a stone,
Nevertheless I have put in a basket
The coins for the ferry.

The Almond-tree in the King James Version

White, yes, pale with the pallor of old timbers,
Thistle-stalks, shells, the extreme pallor of starlight—

It is the almond-tree flourishing,
An image of Age in the Book of Ecclesiastes.

Premonitions, like visitors turning the door-handle,
Cry out, 'It's us. It's only us.'

And I, opening the door from the other side, reply
'Of course. You are expected.'

To memory I say: 'You must be disciplined.'
To hands: 'Do not tremble. Be still.'
To bones: 'Do not ache. Remain flexible.'
To ears: 'Do not be affrighted
It is only the voice of the bird.'

To eyes I say: 'Be faithful. Stay with me.
Do not, looking out of the window, be darkened.'

Yes, it is as I have always been led to believe:
Premonitions, recognitions, the need for acceptance.

The almond-tree shall flourish, and the grass-hopper shall be a burden
It is all in the twelfth chapter of Ecclesiastes.

The Apparition

In a room empty of all but the shift of shadow
And a wooden chest, solid, beneath the window—
What is she looking for there, kneeling before it?

She has lifted the heavy lid against the sill
And with both hands she seems to be dipping, sieving
And letting fall the folds of unseen linen.

I know the curve of the head, the hair gathered
In a sweep to the crown, the long fingers,
The arch of the back and the line of sloping shoulders.

Is it grave or swaddling clothes you are after? Tell me.
Can you forgive me, I ask. What should I have done?
Speak to me, turn your face, give me an answer.

She leaves the linen, shuts down the lid and is gone.
I truly believe that I know her. My distaff side:
My mother, hers, and the long line backwards of women.

Each time I hope to be given absolution.

Michael Dransfield

From *Geography*

III

in the forest, in unexplored
valleys of the sky, are chapels of pure
vision. there even the desolation of space cannot
sorrow you or imprison. i dream of the lucidity of the vacuum,
orders of saints consisting of parts of a rainbow,
identities of wild things / of
what the stars are saying to each other, up there
above the concrete and minimal existences, above
idols and wars and caring. tomorrow
we shall go there, you and your music and the
wind and i, leaving from very strange
stations of the cross, leaving from
high windows and from release,
from clearings
in the forest, the uncharted
uplands of the spirit

Morning Prayer

lord
help us through

another day

who cannot
help ourselves

who may not
make it alone

Geoffrey Dutton

Twelve Sheep

To be still, to be silent, is an act of prayer.

Watering my trees by the red, hard road
I was suddenly aware
That ten yards away twelve sheep were standing,
In a circle, heads in, heads down. So they stood,
Motionless, noses above the red stone.

Are only humans capable of prayer?

Their solemnity was so absurd
I nearly squirted them with the hose.
I was uneasy, as a small boy
On some occasion of great columns
When an echoing laugh would be a comfort.

Those who mumble together do not pray.
Prayers grow like windless trees from silence.

Then one sheep raised his head to stare
Straight at me. And all stood dead still
In the hard centre of the red road.

I remembered an atrocious painting
Of a soft Jesus with a crook, a woolly flock behind him.
'All we like sheep have gone astray.'
'The Lord is my shepherd.' 'Feed my sheep.'

Under the yellow stare of that white-faced sheep
I was forced also to remember
That Jesus was a carpenter.
Whoever dressed him up in shepherd's robes?
I cannot see him whistling a dog in the dust
Bringing in a mob to the yards, the crutching, say,
Pale fingers amidst the maggots and the dags.

The hose still ran. My tree was nearly drowned.
But still that same sheep stared at me
And the other eleven held their circle
Carved in red stone, heads bowed.

And who are we, to be likened unto sheep?
To evade the stare of the murderer, saying
'He acted like a beast.'
To avoid the shamble of youths around the weeping, naked girl.
'They are just a pack of animals.'
To climb a two-legged ladder and look down
On the soldiers shooting children at a range of four yards.
'They behaved like brutes.'
And even in peace, in the hot war of love
To resent the kinship,
The animal coupling, the animal instincts,
To hug the exclusive soul in the act of prayer.

Sheep that go astray eat flowers,
Fan out across the neighbour's crop.
Black sheep drink from the same trough as the white.
In an age of fences there are no shepherds.

The sheep that was watching me turned
And trotted off over the crackling gum-leaves.
That twelfth sheep with his insolent Judas stare.
The eleven good sheep stood stock still
Feeding on air. You would swear
It was in prayer.

‘E’

(Mary E. Fullerton)

Earth

Souls going out,
Souls coming in,
Fill the old earth
With silence and din.

Souls coming in,
Souls going out;
Only their Maker
Knows what it’s about.

Impregnable

Most precious is held from danger,
Walled in a deep defence,
Far from the perilous edge—
The thin circumference.

Marrow, and pith, and kernel
Outlast the hammering storm,
Each in its guarded centre—
A form within a form.

The soul of man made precious
Under reprieve and curse,
Bides for the final heaven
In the sheath of the universe.

Windows

From out his window peering,
In his small jail of being
The lonely prisoner seeks
A more authentic seeing.

As that far Other Window
Flashes a gleam more certain,
Some angry Immortality
Draws quickly down the curtain.

Stephen Edgar

Tenebrae

I

The entrance to Avernus is on Howrah Beach.
Oh yes, I've seen the ground subside
Beneath a she-oak where the sand
Funnelled into a chasm that might reach
Cathedral size somewhere below me, and
A grey sky slide
Downwards like a blanket from a bed.
Like flares above my head

The gulls performed their dying fall by wing and voice,
With a perfect, heartless grace and sweep.
Old shelters, boatsheds had to tell
What always passes in them: LOVE ME BOYS,
SEX, XMAS 76. Why this is hell,
For we can't keep
What we are. Seaweed on wet rocks, its grip
Like clinging fingers, slipped.

II

So what is left? I listen to Roy Goodman sing
Allegri's *Miserere*, a sound
Of disembodied beauty so pure
A Nero would have killed it, envying
A flawlessness he couldn't let endure.
But having found
A heartless perfection, what else could he choose,
Lost in what he'd lose

When his voice broke? Like a light you've focused for too long
The last note clings. I hear the chant
Soothe and perturb the dark devout
When candle after candle is unsung
And a brief voice flares above them and goes out;
For at last they can't
Keep what they are, and will endure the cost
Of being what is lost.

Dead Souls

In the darkness of the kitchen late at night
(All lights have been put out) the notes of Satie
From another room float through like globes

Of a dim illumination. Everything,
Immobile and withdrawn, takes on their sheen,
A glimmer that forbids the inquisitive hand,

A posthumous display, released from the world
Of matter. This shelf of glasses, at eye level,
Glistens like a game of vestigial light,

The ocular equivalent of an echo.
Like spirits of the virtuous dead, they queue,
Absolved from further truck with sense or use.

Out there, on the mountain's slope, a summer fire
Ulcerates the night like a scene from Bosch.
The very silence is conceived of howls.

Diane Fahey

Millipede at an Ashram

Utterly free of mind, it attains
the centre of the meditating circle,
begins a series of yoga asanas—
curving, contracting, that pliable
form adopting with ease the postures
we strain for. At rest it becomes
a spiral of self-knowledge, a black
uroboros with silver spine.

One of us
not yet touched by enlightenment
flings the millipede out the window.
We stare into a hollow core of space
again, remember to keep breathing.
On a sunlit leaf, the millipede
nibbles, does nothing.

Robert D. FitzGerald

From *The Greater Apollo*

IV

I look no more for gods among
the lace-like ferns and twisted boughs,
and little care though these should house
the riddles whence all life has sprung:
the Presence whose long dream endows
the universes with the store
of all that they possess, or fays
dancing light-foot down leafy ways,
or Pan, Jehovah, Brahma, Thor—
the man-made gods of earlier days.

Not now, as once, I seek the cause,
the soul of life, the secret tide
whereon all things we know of ride
obedient to destined laws.
I seek no longer to divide
being from that which gives it place,
matter from spirit, for I rate
animate and inanimate,
life, time and substance, action, space,
as oneness infinitely great.

But I grow angry with my thought
which plunges rudderless and amiss
within this stormy, deep abyss—
presumptuous atom to have sought
(embodiment of all that is)
the Greater Apollo whose wide reach
embraces the unguessed-at years,
shaming our ethics, creeds, and fears,
shaming the priests for what they teach,
shaming the narrow-visioned seers!

The valley path is calm and cool
as I walk here between green walls—
and those are diamond waterfalls;
this is a bird; and that's a pool.
I heard their far insistent calls
and gladly have returned to these.

What is revealed to me and known
beyond material things alone?
It is enough that trees are trees,
that earth is earth and stone is stone.

Revelation

This clue to their God you may disclose
even to the faithful: in all that flows,
what it flows from and what towards.
Most else is but a threshing of words
flailed out so finely in pursuit
of the unsiftably minute
that you lose truth under dust of talk.
Here is the grain of it in the stalk.

From *Eleven Compositions: Roadside*

VI

Never the same tide returns
to its old course of flesh and bone:
daily the living man discerns
how all flows through him on and on.

This has been said: that at your death
your struggling force ebbs back to God
as to some storage-dam of breath
for those who follow where you trod.

But no; I saw the tide that fell
from gasping mud and flabby weed
and rocks of grating oyster-shell;
and though life came again indeed

with the next current, what then rose
was a new water, feeling, blind,
for this calm reach, not as it chose
but pressed and jostled from behind.

You'll say each particle had learned
lately the path it rolled along
where the retreating ebb had churned,
a routed host, a jumbled throng;

whence came this thought of life being drawn
from a mixed pool of earlier lives;
but water's neither beads nor spawn,
and not as fragments birth arrives,

but whole and single from that vast,
the self-renewed, be it God or sea.
Life's not what bubbles and goes past
but flow itself, through time, through me.

From *Insight: Six Versions*

Creak of the Crow

Hardship and hazard
led to holy land.
The skull-dwelling lizard
bakes there on sand;

and nothing can atone
for loss that must fall
of the black stone
from the white wall—

black still and square
in true believers' eyes;
firm—though Mecca's bare
for us, the unwise,

who turn from mosque and mullah
to whirl and upthrust
of the voice of Allah
in a spire of dust,

and fastening faith
beyond works that show
might dream the world wraith
but for creak of the crow.

John Forbes

Ode to Doubt

Luminous substrate
our views are just a veneer on, or like
the slats in a venetian blind
changing with the weather or the mood I'm in—
without you the light drones behind our heads
and we know what the world is like
or better be, if it wants to avoid
the pious grindstone of our self-regard
and tho' certainty has its saints,
to just step outside with you, like drinking
considered as experience,
would make their good deeds nobler
                    and, if nothing else,
abolish hagiography as a genre
(not to mention theories of reflection
and the clear-eyed luxury of teenage beliefs—
America's tribute to your absence).
We stand to be corrected & sit down unassuaged
because of you. You keep us modest
like the stones I used to shift around as a kid
assuming they needed a change of view. That
was the closest I got to playing God,
as if each brick had a message wrapped around it
or, vice-versa, the world was a fake parcel
                    with each layer of tissue
gauzy, ornate & gorgeous, keeping us from you.
Blanchot says, 'at the moment of death
we each experience our life as a lie', so
no wonder dying is important to believers.
With you that shade is always there & we become
a happy extension of the rocks & trees,
                              no longer certain
but a part of what we know.

John Foulcher

Wars of Imperialism

From 66 to 70 AD Rome and Palestine tore
Jerusalem—in history it's small, strong only to the Jews
who lost, living since then in tents.

Yet I think we should see more in this:
things ended when their Temple fell. Like a honeycomb,
cells progressively damp and alone, the great home
of God was peeled; brick by brick,
places more holy exposed to the light, the Romans marched,
spiralling in. And when light came
to the last place, the centre, He must have left . . .

Did they know what they set free then?
Something was there,
changing to light from ages of nothing—
scorching the known world, ripping the new world.

Reading Josephus

Once, reading Josephus,
I found this description of Christ: he was a black man,
very nearly black,
tarred with the Palestinian sun
and shorter than most.
His hair was never cut. His nose beaked over,
farcically Jewish.

Hunchbacked
as well, a haversack
of gristle and meat
lugged about, pressing his spine down;
eyes tilting to the sand before him. Imagine that. Those
hefty wooden verbs
dragged out and thrown
before the listeners—
not sublime at all, not the easy construction of a man
nailed upright.

This was a lame Saviour,
glazed with sweat, heart pounding from the body's haul
up to Calvary,
where his tall disciples
and the squat metal guards
had to bend back their necks to see him
hammered out straight at last. Ascending,
with all the pretty angels.

Barbara Giles

Beauty for Ashes

Fair promise, this? The ashes of life exchanged
for the beauty of holiness? I read the guides,
all admirable, but I'm too stiff to kneel.

The Way appeals, appals and even Quakers
make a profit. Existentialists,
saints without God? The old sins itch.

I open the stoppered jar for just one sniff.
The ancient demons all come tumbling out,
unkinder, bolder, shrieking all night long.

Then it makes sense to fiddle while Rome's burning,
to end in ashes and no beauty left?
Incline my ear to hear a truer music.

James Gleeson

Drill of Central Thunder Notes: the city expects Christ:

the alpaca night shivers with the suddenly passing light;
codes of more demanding, lightless times,
the falling rain, white, in the off chance:
city is in uproar: all about:
hysteric full-squares driven to by the wind
waves: wet leaves, one on the cold throat
one slithers over my thigh   others
pattern my breast: make high sounds:
rain-cold fence of wind
bringing me the wetness of leaves
the emotion of crowds in the street the feel of people
the people themselves feeling, the one mind
the many bodies the many bodies the oneness
the division the singleness the many
the clog of phrase, stick of the trite
the urging, the marching: the march the wind under our feet
our cold armpits: shrunken sex: small: wind
in our heads through the bones of our fingers
the finishing coldness the unfinished eyes the finish
the final coldness the coldness of extinction
the ending body: many: the ugliness of similitude
the same the monotony of the unit
who lives in the five great cities?   to watch
the star-eggs hatch in disaster   to watch
the fine nadir   would?   here:
here, in the 5 a.m. of our times
would go apart in the unending night-time,
known that the night cannot be ended.

say something? in the loose corners of the wind:

in a moment we looked back
and saw god-ourselves,
or what passes.

Peter Goldsworthy

From *Mass for the Middle-Aged*

Libera Me

Deliver me, Lord, from the threat
of heaven, from becoming the angel
who is not me, who smiles
faintly, fondly

before shrugging me off
like some stiff, quaint pupal case:
the battered leather jacket of the flesh,
evidence of misspent youth.

Grant me, Lord, this last request:
to wear bikie colours in heaven,
a grub among the butterflies.
And this: to take all memories with me,

all memories that *are* me,
intact, seized first
like snapshot albums
from a burning house.

Answer, Lord, these prayers,
for I would rather
be nothing
than improved.

From *A Brief Introduction to Philosophy*

What Comes Next

There is nothing as empty as the future,
or as bleached and pale blue: a type of summer,
a long school-holiday, unpunctuated even
by our little lives, rounded with brackets.

Outside those brackets, what? Or—far worse—why?
Don't ask so many questions wise adults
repeated, often, when I was young—
but each year I push an extra candle

through the crust and panic:
another pilot-flame to extinguish, quickly,
lest something uncontrollable ignites,
or I find myself breaking through icing

into the molten stuff beneath, suddenly
reduced to composite materials.
Perhaps this is the final homecoming:
a fair and even redistribution of matter,

my atoms permitted to cease their restless
jiggling, at peace among the other particles;
my bits and pieces returned to where
I sprang from—or less I, than me:

and less me, than him: his handful of carbon
returned to that topsoil, his water-quota—fifty litres—
to those streams and clouds, his ash to that ash;
his dust to *that* dust, there, no longer mine.

Alan Gould

The Henty River

This river is so still its greens connive
to move without a change of place. It holds
the forest and the forest holds the river.

We've stopped between the towns of Strahan and Zeehan
to roll a cigarette and let the stillness
take our lives from time. This is not August,

this is not afternoon. This is green,
meaning river, stillness meaning forest,
and each is different and is the same.

It's utterly unhuman, like paradise.
To leave this bridge and enter trees or water
is movement we can make if we become

molecules or the quality of green;
this is now happening. The silence is
as deep as memory. The stillness is

as whole as time. Strahan and Zeehan are now
remote as what might have happened. We've seen our home,
unearthly calm, the greens of deepest yearning,

and may not do so in our lives again.

Robert Gray

Dharma Vehicle

1

Out of the reach of voices
in the wind.

Camping at a fibro shack
fishermen use—
swept with tea-tree branches, and washed down
with kerosene tins of
tank water.

Like banners raised,
all these eucalyptus saplings—
the straight trees.

A sea-breeze
over the grass headland, where fallen, white
branches swim;
leaves here
are shaken all the time,
shoes that run
on stone.

My bed
a pile of cut fern.

*

And the Pacific Ocean mornings
in windows rinsed with
wet handkerchief,
among the whitish-grey, ragged paperbarks—

that glint
all over,
in long shoals,
of translucent
scales.

At night,
lying by the fire
outdoors—seeming to lean above
moon and stars afloat.

The distant cannon
of the waves . . .

The paperbarks climb
slowly,
and are spreading out, like incense-smoke.

*

I read beneath the trees all day,
caught-up
with those old Chinese
who sought the right way to live, and found
one must adapt to nature,
to what is
outside our egotism;
who loved this earth.

'Here I am
gulping the stuff from the fountain
and willing to let it (Lucretius)
trickle out of my mouth.'

In India, the Buddhists
praised insensibility
to the world
('Doth not the Hindoo
lust after vacuity?');
but with Buddhism's arrival in China,
by the T'ang,
in the time of Hui-neng, the sixth
patriarch, there'd come
a complete reversal of such *dharma*—

There is the Other Shore,
it is here.

*

It is not reaching into any deep centre,
but to awaken the mind without fixing it anywhere.

A man who goes into trance
and has no thought or feeling (Hui-neng)
surely is no better than a block of stone
or bit of wood.

But to know pleasure as pleasure
and pain as pain
and to keep the mind free from all attachment
is what's called No-Thought.

*

I turn out the lamp.
Leaves, twigs, berries falling
on the tin like rain
in the night.
—It was the monk
Fa Ch'an-ang, in China,
dying,
heard a squirrel screech
out on the moon-wet tiles, and who told them
'It's only this.'

*

Only this.
A wide flat banana leaf,
wet green,
unbroken, leaning on
the glass.

The mother-of-pearl of a cloudy dawn.

2

How shall one continue
to confront every morning
this same face in the mirror?

Anxiously peering,
demanding—
such intolerable self-pity;

hysterical, and without decency.
Impossible marriage
with such a face, that eats up other people.

I do not want to be this sort of cripple
in the world any longer;
not for any of my excuses for being

to remain,
not for any of my possibilities.
I do not want to be what I am.

I'm woken here,
I would like simply to walk away.
And live without saying that I live,

without me
as the filament, the grains, the sedimentary content,
the matter to be taken into account.

And continue,
but without this continuing; certainly,
not to remain defender of such a proposition,

which, every next moment, life is going to contradict,
and with the back of its hand,
and with its fist.

*

When you are suffering
and you want to be free
of that which torments you,

it is not greed,
is it?
This is something more basic than

the calculations of thought.
And this is why I've felt
it's possible

to elude the mind,
whose confusion has continued
for too long.

The summer's almost gone.

3

The holymen whom Gautama sought-out in the forest—
torment of a leper—
knew about Transience,
as did Heraclitus, about this time,
but taught there is a soul,
Atman,
'I',
and that it's the same as the World-Soul, Brahman,
of abiding nature.
Gautama saw there is no cure for the Self
in such belief.

'I saw the thorn
that is piercing to the heart of men'—
and belief in the soul is part of its poison—
the thorn
is one's subjective desire,
to which a man clings above all,
tenacious like the shark

and as cruel—
'If this thorn is drawn out
one is calm and knows peace.'

He could not find his relief among those *sannyasin*,
and so, went on alone;
staring in
where all the other's knees had failed them—
on the edge of the buffalo pasture
in blue smoke of moonlight,
in the wet grass,
with mud running on his body;
or among tree roots,
and there he saw the ragged wild flight
of the stars,
a particular night—
it was like,
as many others have said, since then,
'the bottom falling out
of the washtub',
or
'like flowers suddenly blooming
on withered trees'.

*

'No God, no soul'—
It is all like a mountain river,
travelling very far, and very swiftly;
not for a moment does it cease to flow.
One thing disappears and determines what is arising,
and there is no unchanging substance
through all of this,
nothing to call permanent,
only Change.
That which is the substance of things
abides as nothing
and has nowhere
a nature of its own.
Its essential nature is Nothingness.

*

In Western thought, this recalls something that Engels said
in 'Dialectics of Nature'—
that 'motion is the mode of existence
of matter';

there is no form of matter that isn't in transformation
and therefore
no form that's an essence.
'Matter, as such',
Engels wrote, 'is the pure creation of thought . . .
an abstraction';
matter only exists in particular forms.

So that these transient things, themselves, are what is Absolute;
these things
beneath the hand, and before the eye—
the wattle
lying on the wooden trestle,
pencils, some crockery,
books and papers, a river stone,
the dead flies and cobwebs
in the rusty gauze.

4

I am woken here when 'the sun gets to its feet
shouting'.
The sun takes a stride,
'wearing its waistband of human hair'.
I go out, over the morning's copious small water,
never touched,
and the golden breath covers the dense forest and the mountains,
the paddocks below
that are streaked with dead trees.

I walk down a long slope
where the bush is cut far back on either side;
the early sky, so light,
has a feeling of
the first day up again after illness;
the dew is dashed in the grass,
blue,
gold, red; as you pass above,
it lights up, everywhere you walk.

From this hillside
I can see, around a solid, wind-levelled, slant mass of trees,
the ocean—
like silver foil
that's been crumpled and smoothed again.

And below me, dark timber,
with those topmost cauliflower-clumps of eucalyptus
scattered, opaque,
against the ocean light.
Beyond, there are banana plantations
right along, over the billowing hillslopes,
and a few tin roofs
lit-up like dangling water-drops.
I hear, faintly, a dog yap,
and can see blue smoke
that is staggering along the air a little way.

I go further down;
wade a lagoon of whisky-coloured grass
onto the dirt road (soaked
to the knees), and pass
a deserted schoolhouse, with its red iron roof and tank,
and tennis court
lying within wire netting; and now a flock
of parakeets
sweeps by—it banks
on the morning, dark lift of wings—is settling
everywhere.
They're leaping about amongst the trees, and some dip into the
court,
vague behind wire
as if flying through mist—
wheel up again
catching the sun—their feathers then
the colours of that dew . . .

5

It was in China that men first could say
of this transitory world
it is Nirvana.

The Taoists had seen the universe is Self-Existent,
and that all particular things
spontaneously arise.

'If Heaven had produced its creatures on purpose
it would have taught them to love (Wang Ch'ung)
and not to prey upon each other.

Rather, all things have come about through Transformation,
because they are one.
You do not find anything superior to things.'

Such a universe was spoken of as a Great Furnace
in which all that is
shall burn.

It is a fire that consumes the fire, undiminished.
How could Heaven have pity for that fate
its nature brings about?

'Though all is in destruction and regeneration at once
there is tranquillity in this disturbance. (Chuang-tzu)
Tranquillity in disturbance is called Perfection.

There are ten thousand things being transformed,
and the sage is transformed along with them
without difference, without end.

Therefore, his movements are effortless as water;
his stillness, deep like a mirror;
his response, an echo.

His rarefied condition makes him seem to disappear.
He accepts his body with pleasure,
forgetting life and death.

To him there is nothing in the world that is greater
than the tip of a hair
that grows in spring.'

*

When something comes into existence
it is because of conditions that are favouring—

all-that-is, being interdependent,
combines to bring it forth,

and thus nature is good to man,
or at least, is more favourable than against him.

And everything appears, the Taoists said,
in dialectical relation—

it is like two stags
that lock horns close to the ground—a sound

of bamboos knocking together—
whose playful blood grows erect

in their veins: pushing
they manoeuvre

and stagger,
the dust arises

as all of these floating worlds.
On each world

and in every event of each
the same two stags contest.

And Lao-tzu, on leaving the Empire of Han,
at some vast age, to die—

riding on his buffalo
that was like a torrential rain—

wrote a poem to the people, and left it with the border guard:
that men should confess

it is the opposite of what we love
is good to us,

that it's only this weeping
which can make us glad.

*

The things of the earth
fill men with life
and swarm, like red corpuscles to a wound,
to do them good.
The earth feeds men aright;
the five grains are to feed them,
and the beans,
and the leaves for their soup;
and the water the same,
it goes down alive
inside men; and these fruits,
they feed men well.
And even Death—
the vinegar
that is found in the dish.
One ought to go out
into the forest and sun,
bathe in the streams and ocean,
and care for the body with oils
and comb the hair and decorate oneself
to sleep with another,
and join one's friends
in the grove of summer,
or beneath wide eaves
in dark weather, when the rain drips,
bringing wine.

And wander along the mountainside alone.
Because life is fleeting,
it is the breath of a bull in the wintry dusk.
Throwing away the self, 'let us hasten
to enjoy this life'.

6

The image of Buddha became a fat Chinaman
who was rolling on his haunches in a fallen-down robe,
twiddling as a fan
the end of a banana leaf,
with tits that wouldn't have looked out of place on a sow
and a laugh like a slice of watermelon—
to tear-up the conceptions of the mind.

Onto that way in harmony with nature there was joined
the sharp means of Release.
Before this, the Taoists were content with passivity,
and for discarding the self, most often
made do with wine.
The Emperor Wu-ti was first to hear the Unique Insight,
who asked of Bodhidharma,
define Buddhism.
Bodhidharma replied: Vast Emptiness.

*

Ma-chu got up onto the wood platform,
eased his legs in the Lotus,
laid aside his fan;
he started to trail smoking water on the green tea powder,
beating it with a whisk,
and looked over wet gravel, the heads of all the assembly,
between the darkness of wide, heavy doors,
to a lemon colour in the garden;
then he said to them, 'There is no Buddhahood for you to attain;
cling to nothing, that is the Tao',
and signalled for the crack
of the woodblocks together, for them to leave;
and sipped from the bowl, alone.

*

And there was a Master, Hsuan-chien,
told his students, after they'd sat in the courtyard for many days,
prostrating themselves, to be taken in,
'Pull on your clothes of a morning

and work along the hillside with the others,
or rake the leaves,
until you hear the dinner drum;
eat your meals,
and go to the john when you have to—
That's all.
There's no transmigration for you to fear, no
Nirvana to achieve.
Just respond to all things
without getting caught—
Don't even hold on to your Non-Seeking as right.
There is no other wisdom to attain.'

*

An afternoon rain
is drifting like sails of smoke
among these paperbark trees, about the shack;
it is the crumpling sound of cleaning up
cellophane wrap, close by,
and in one place, that green slap
of cow water.
'Sit straight, in Padmāsana, like a mountain,
keeping count of the breath, over and over,
so as not to touch your thought,
the eyes left open.'
You cannot dwell anywhere.
Realizing, beyond the intellect, that 'I' do not exist.

There is a soft, wet, vivid green,
a paddock, that rises
full as a breaker's first lifting,
now, after the cold stream;
and above it, struck by watery sun,
scaffolding tree-trunks, branch-beams, obliques, that shine
whitely, out of the black cumulus of bush.
And I hear rain tapping,
as if on canvas, from the guttering, and from where
a bird has skittered
all about this window, shaking the wet tree.

'A mind that's like a mirror,
in which things pass and leave no stain.'

7

I'm coming back with a haversack from the shop;
a beach resort
miles off,
walking all the way at the water's edge
along the empty sand.

Those shops they have forgot
to wind.

The one street, an old faded print,
squints in the glare;
outshone
by the plate-glass sea.

And I climb, one after the other, over
the headlands, on rock,
in late afternoon,

looking out to where all the clambering, wilted,
flaring

Ocean

begins, of a sudden, its bellowing and stamping,
the lowering of its shoulders,
a smoke-spray
blowing from them.

The surf comes in as though alive and tearing free
from under the net of foam—
making its break
with the panicky, bounding gallop of some great animal up
hopeless
onto the slippery shore.

And all the time along the horizon
those clouds,
that are like mountains with cliffs and valleys, now,
in the last, stretched-out sun—
that dreaming, far-off,
impossible land.

Night comes
quickly, over the water, as if water
flowing into the space left by the withdrawing sun,
and foam spreads
flatly all around me, phosphorescent,
bubbling and crackling in crab-holes in the sand.

The waves flicker
like a book left in some vast, empty house,
to a noise of doors slamming.

I am weary and cold, by now.
No one about.
Only, across the rising moon's long beam,
a bird flies,
skimming the horns of the sea.

This long beach,
beneath the immense imagery of night
and the night-bird's croak,
keeps on disappearing into the mist and dark.

And at such times, 'even in the mind of the enlightened
there arises sorrow',
so it's all right.

Lesbia Harford

Summer Lightning

Just now, as warm day faded from our sight,
Hosts of archangels, fleet
On lightning-wingéd feet
Passed by, all glimmering in the busy night.

Sweet angels, bring no blinding truth to birth,
Give us no messages
From heavenly palaces;
Leave us our dark trees and our starlight earth.

Buddha in the Workroom

Sometimes the skirts I push through my machine
Spread circlewise, strong petalled lobe on lobe,
And look for the rapt moment of a dream
Like Buddha's robe.

And I, caught up out of the workroom's stir
Into the silence of a different scheme,
Dream, in a sun-dark, templed otherwhere
His alien dream.

A Deity

Sometimes I think God has his days
For being friends.
He says: 'Forgive my careless ways.
No one pretends
I'm always kind; but for today
Do let's be friends.'

And grudgingly I make reply,
'Nice sort of friends.
I think it's time you had a try
To make amends
For things you've done; but after all
Suppose we're friends.'

I am no mystic

I am no mystic. All the ways of God
Are dark to me.
I know not if he lived or if he died
In agony.

My every act has reference to man.
Some human need
Of this one, or of that, or of myself
Inspires the deed.

But when I hear the Angelus, I say
A Latin prayer
Hoping the dim incanted words may shine
Some way, somewhere.

Words and a will may work upon my mind
Till ethics turn
To that transcendent mystic love with which
The Seraphim burn.

A Prayer to Saint Rosa

When I am so worn out I cannot sleep
And yet I know I have to work next day
Or lose my job, I sometimes have recourse
To one long dead, who listens when I pray.

I ask Saint Rose of Lima for the sleep
She went without, three hundred years ago
When, lying on thorns and heaps of broken sherd,
She talked with God and made a heaven so.

Then speedily that most compassionate Saint
Comes with her gift of deep oblivious hours,
Treasured for centuries in nocturnal space
And heavy with the scent of Lima's flowers.

Charles Harpur

The Silence of Faith

A thousand million souls arise
  Out of the cradle of To-day,
And, like a storm, beneath the skies
  Go thundering on their destined way!
    But ere To-morrow's sun
    His ancient round hath run,
  The storm is past—and Where are they?
  Is asked of Faith by pale Dismay:
      O say,
    Where—where are they?
And faith—even Faith herself, hath not a word to say.
  With her serene assurance thrown,
  Like moonlight, into the Unknown,
  And all her clasping tendrils curled
About the steadfast pillars of the never-failing world,
  To that wild question of Dismay
  —Yet hath she not a word to say:
  And only lifts her patient eyes
    Up from earth's tempest-trampled sod,
  To fix them,—out in the eternal skies,—
    On all she knoweth—God.

Robert Harris

Ray

It seemed wise, wise. To get it wrong
about His ray. Trying to plunder Christ,
for I heard the appalled hush at His name

in a bookshop, during a poetry reading.
Soon He was calling, not He without His Friend.
In from behind the winter wind.

The loudest rain could not drown
that soft knock. If then I heard words
they were, Why not come from hiding?

You're an archetype, I flung back. So
go away. Or said, Nah. Listen, says Christ,
listen be deaf you are deaf now you aren't,

listen. I will be back. Meantime keep
that wisdom. It helps you. Do what? Oh,
die. Or if you listen, to hear the lie

under each old friend's warning.
The old, habitual, single Lie. To know you
are blind, now in My light, go seeing . . .

The Call

Still, I lay awake in the dark . . .
I thought pretty lusts had some ugly results
and that the world's bright trash was occult.

And carnage steals us from the eye of summer,
we cannot explain; but sometimes
the cockatoos are upset,
the flock miss one of their number.

And still I lay awake in the dark,
that was where I would have to fall—
out there somewhere in consequences,

in a desert become too hostile for survival;
if by weapons, disease, or auto design,
a cipher subjected to die for enormous grudges.

A cigarette answered that I was alive
though it seemed I had waited for centuries
becoming sure that the dark itself was active.

I was counting too a propensity to misfortune
back to great-grandfather's funny ways.
He worshipped a spurious architect of nothing

and walked on his hands on ordinary Sundays.
But He who drained the cup once for all
and did so under hell's lowest stair,

Christ, called me through from the other side of lightning.
Now I would seek out a comelier praise;
then I felt like one in a room of crimes

as the blind rattles up, and the light crashes in.

The Cloud Passes Over

The high spring winds
arrive unannounced,
sparks bowl superbly
and dangerously
as heaving power cables rub,
cars lurch
on the mountain highway,
children and dogs
are restless,
water flows sideways
from faucets outdoors.

What happens to things,
to a roof
or municipal flowers,
bending as though they were
voluntarily rifled,
had blossomed for wind
to lift and touch and know them
gets lost for days—
in the roaring among young eucalypts,
the uproar
threshing cypresses,
wobbling speed signs,
the glass shaken out
from a wretched bed of dried putty
to shatter as a week of conversation . . .

Some nights
　　　　the Lord God of waters
moves down the freshwater,
　　　　the estuary, rivers
veiled in darkness.
　　　　In silence He inspects
the snags
　　　　where the bank drops away,
examining every rotting trunk,
　　　　every hole where fish sleep.
He sets aside mullet and trout
　　　　for koori people,
for dairymen mourning
　　　　under the quota system.

But these nights
　　　　there aren't any fishermen out
from caravan and tent enclaves,
　　　　their hair on end,
their lines frightened in;
　　　　no little white cloud
with damaged oars
　　　　passing over so carefully
that nothing below
　　　　may hear it think.
The Lord of all
　　　　is at large throughout His Creation.

Mornings they find
　　　　in a boarding hotel courtyard
His gale has raised
　　　　from a tree a branch,
red flowers
　　　　spattered,
blowing around the white pickets
　　　　and every lifted grass-blade
cranes, on tiptoe.
　　　　Old men with catarrh,
the kitchenhand who can make
　　　　them back steps glitter
find the slat and steel
　　　　chairs flat-a-back.
These too appear to
　　　　wonder
up, through sky and through spring,
　　　　through every seemingly
physical thing.

Isaiah by Kerosene Lantern Light

This voice an older friend has kept
to patronise the single name he swears by
saying aha, aha to me.

The heresy hunter, sifting these lines
another shrieks through serapax and heroin
that we have a culture.

These are the very same who shall wait
for plainer faces after they've glutted on beauty,
a mild people back from the dead

shall speak the doors down
to the last hullo reaching the last crooked hutch
in forest or forest-like deeps of the town.

These who teach with the fingers and answer
with laughter, with anger, shall be in derision
and the waiting long, and the blue and white days

like a grave in a senseless universe.
I believe this wick and this open book
in the light's oval, and I disbelieve

everything this generation has told me.

The Eagle

All day the telephones rasp
and nothing in the city can
discern you in its mother tongue,
if you are alone
or at your desk,
dark haired, a lily of Solomon,
walking in grace at seventeen.

Not the telephone, the intercom,
or the electronic Adler,
the cabinets and birthday cards
and cafe-bar beside the blinds
which horizontalise the view.
Nothing will do,
though the suburbs stretch to sunset,

nothing will prove a substitute
for clearer air, though all day
the typewriter chatters, but at
the self's fringes
you feel the insistent, slight pressure
lifting you from earth to prayer.
There, in His praises,

the eagle rests on an updraft,
as He dwells above time in His praises,
past and future. And those who pray
with you know also that each heart's a desert,
flesh is as grass, its joy as a wildflower.
But the eagle rests, a freed spirit,
and the lilies rise with the river.

Kevin Hart

Master of energy and silence

Master of energy and silence
    Embracer of Contradictions
Who withdraws behind death
    Like horizons we never touch
Who can be One and Many
    Like light refracting through glass,

Stepping in and out of logic
    Like a child unsure of the sea
In and out of time
    Like an old man dozing, waking,
In and out of history
    Like a needle through cloth,

Who we chase and bother with theories
    Who hides in equations and wind
Who is constant as the speed of light
    Who stretches over the Empty Place
Who hangs the Earth upon Nothing
    Who strikes like lightning.

The Stone's Prayer

Father, I praise you
For the wideness of this your Earth, and for the sky
Arched forever over me,
For the sharp rain and the scraping wind
That have carved me from the mountain
And made me smooth as a child's face.

Accept my praise
For my colour, a starless night,
That my width is that between the first two stars of evening
Reflected in the water,
That my quartz flashes like lightning
And reflects the glory of your Creation,

That you have seen fit
To place me near a stream and thus to contemplate
The passing of time;

For all that is around me I sing your praise,
For the fierce concentration of ants, their laws,
For all that they tell me about you.

Keep me, I pray, whole,
Unlike the terrible dust and pieces of bone
Cast about in the wind's great breath, unlike men
Who must suffer change,
Their endless footprints deep as graves;

Keep me in truth, in solitude,
Until the day when you will burst into my heavy soul
And I will shout your name.

Approaching Sleep

Footsteps in the attic, the crooked sounds
You hear at night, the train's blind whistle or
Dead letters slipped beneath your bedroom door,

And still there is your heart that beats upon
Your ear and fills you as you lie in bed;
It beats and beats but cannot keep good time

And lets it drip like water from a tap.
You write a letter of complaint to God
While half asleep, forgetting the address.

Outside, the night is wide as a winter lake
After the heavy rains, and it is June
With days that open like a Chinese Box.

If anything is real it is the mind
Approaching sleep, listing the tiny bones
Within the ear: *anvil, stirrup, hammer* . . .

The surgeon placed them on a woman's watch,
The seconds crudely sweeping underneath.
Within the ear, a fine Dutch miniature

With cool canals, a blacksmith by his horse,
A small boy playing on a smaller drum,
Old women who darn their shadows again each dusk.

There is a monster in the labyrinth
But always behind you, walking when you walk:
It is too late to get out now, the watch

You hold up to your ear stopped long ago;
That angry letter you wrote to God returns
Addressed to you, but now means something else.

Facing the Pacific at Night

Driving east, in the darkness between two stars
Or between two thoughts, you reach the greatest ocean,
That cold expanse the rain can never net,

And driving east, you are a child again—
The web of names is brushed aside from things.
The ocean's name is quietly washed away

Revealing the thing itself, an energy,
An elemental life flashing in starlight.
No word can shrink it down to fit the mind,

It is already there, between two thoughts,
The darkness in which you travel and arrive,
The nameless one, the surname of all things.

The ocean slowly rocks from side to side,
A child itself, asleep in its bed of rocks,
No parent there to wake it from a dream,

To draw the ancient gods between the stars.
You stand upon the cliff, no longer cold,
And you are weightless, back before the thrust

And rush of birth when beards of blood are grown;
Or outside time, as though you had just died
To birth and death, no name to hide behind,

No name to splay the world or burn it whole.
The ocean quietly moves within your ear
And flashes in your eyes: the silent place

Outside the world we know is here and now,
Between two thoughts, a child that does not grow,
A silence undressing words, a nameless love.

The Gift

One day the gift arrives—outside your door,
Left on a windowsill, inside the mailbox,
Or in the hallway, far too large to lift.

Your postman shrugs his shoulders, the police
Consult a statute, and the cat miaows.
No name, no signature, and no address,

Only, 'To you, my dearest one, my all . . .'
One day it fits snugly in your pocket,
Then fills the backyard like afternoon in Spring.

Monday morning, and it's there at work—
Already ahead of you, or left behind
Amongst the papers, files and photographs;

And were there lipstick smudges down the side
Or have they just appeared? What a headache!
And worse, people have begun to talk:

'You lucky thing!' they say, or roll their eyes.
Nights find you combing the directory
(A glass of straw-coloured wine upon the desk)

Still hoping to chance on a forgotten name.
Yet mornings see you happier than before—
After all, the gift has set you up for life.

Impossible to tell, now, what was given
And what was not: slivers of rain on the window,
Those gold-tooled *Oeuvres* of Diderot on the shelf,

The strawberry dreaming in a champagne flute—
Were they part of the gift or something else?
Or is the gift still coming, on its way?

William Hart-Smith

From *Christopher Columbus*

Psalm for Himself

To sit between kings, to earn the right
to kneel and be raised;
to kiss the jewelled fingers of God's vice-regents.

For in serving kings do we serve God;
and our affection for them
is directed unto Him,

by whom all persons are maintained,
the humble and the mighty,
during all their lives, each in his state.

And some in serving the kings of the earth,
and being, according to the Holy Scriptures,
the chosen servants of the Lord,

are directed to make discoveries
of new lands, to subdue them
and establish among the heathen the knowledge of God;

to gather great riches, in order to do great works;
to recover the Holy Sepulchre at Jerusalem
from the power of the Infidel.

Therefore a hunger for gold is a hunger for grace:
great things are done by gold:
by it a humble weaver of Genoa

is raised in virtue
to sit in glory and honour among the mighty
for ever and for ever.

Negative

Pointed my camera at heaven
to photograph God many times
by daylight flashlight flood
with focus set always at infinity

juggled a bit with aperture and angle
finding my arms trembled
found the tripod useful only
busy clicking the shutter I completely

forgot to wind the spool so that I have
the weirdest abstract a sort of
palimpsest hardly worth the printing
God has so many many faces.

Aaron's Rod

Aaron unties the sack
fastened at the neck
with cord and spills a snake

on the ground at Pharaoh's feet.
In the sack supine and slack;
on the ground turns back to strike,

lunges at Aaron's heel.
Aaron rolls his sleeve
and deftly steps aside,

takes it by the tail,
lifts it from the ground,
the snake becomes a staff

stiff, helpless, rigid;
the shame of its underside
pale and barred with frets

like the neck of a mandolin.
Opens the sack again
and thrusts the snake within.

Thus snake becomes a rod
and back again to snake
all for the glory of God.

Ambrosia

He lifted a drop of ambrosia
on a length of brittle straw

and let the bead of nectar run
back along the straw

towards His hand.
Lifted the straw

and tilted it towards the earth again until where the droplet gathered

seven colours flashed.
With it he touched

a creature's being,
the creature of all creatures

that he most loved and treasured.
One assumes

the quantity was not precisely measured.

Gwen Harwood

A Case

Uprights undid her: spires and trees.
One night she lived a vital dream.
By water and by land she came
delayed by manifold stupidities
into a wicked, feasting town.
Her Samson mind cracked pillars down
and left no trees, no upright towers.
By righteousness endowed with powers

extravagant beyond belief
she resurrected from the gutter
the President of Dogs, whose utter
gratitude made words of barks: 'O Chief
Lover of Cleanliness, no more,
I swear, shall dogs befoul your door
or copulate in public places.'
She resurrected girls whose faces

purified of alizarin shame
were safely quarantined from sex;
charms against men hung from their necks
to the division she would never name.
All sweet, all clean this level town.
A phallus rose, she whipped it down.
Day broke.
                    Erect, the bawdy spires
poked in red clouds' immodest fires.

She bathed. She munched her food chopped raw.
Blackstrap molasses charged her power.
'Shadowy Redeemer, come this hour!
Help me enforce thy horizontal law,
and scourge the crude obscenities
of dogs and girls and posturing trees.'

She met him in a crowded street,
tore off her clothes, and kissed his feet.

Home of Mercy

By two and two the ruined girls are walking
at the neat margin of the convent grass
into the chapel, counted as they pass
by an old nun who silences their talking.

They smooth with roughened hands the clumsy dress
that hides their ripening bodies. Memories burn
like incense as towards plaster saints they turn
faces of mischievous children in distress.

They kneel: time for the spirit to begin
with prayer its sad recourse to dream and flight
from their intolerable weekday rigour.
Each morning they will launder, for their sin,
sheets soiled by other bodies, and at night
angels will wrestle them with brutish vigour.

Midnight Mass, Janitzio

In darkness, earlier gods preside.
A child with turquoise bracelets lays
a plaster child in straw. Beside
the virgin hangs a quetzal plume.
The congregation stands and sways
with delicate rhythmic grace. The gloom

of the high nave is ribbed with gold.
Beneath the Stations, in fresh paint,
are symbols of the gods too old
to be forgotten: serpent's eye,
rain-spirit's mask. A wooden saint
with real teeth and hair stands by

the dancers' feathered drums. Tonight
the Lord is born, and flickering strings
of candles sheathed in paper light
streets where in every doorway hangs
a tinsel star, till morning brings
hunger and sickness, gods with fangs.

Revival Rally

A delirium of shapes rising and falling:
bodies shake with salvation, arms fling wild
in appeal to the ceiling of the Gospel Hall.
A cripple throws her crutch away, revealing
a nightmare leg. A mother lifts her child,
which howls in terror at the preacher's blessing.
The wicked pianist leaps up, confessing,
well-rehearsed sins; sits down; integrates all

the abundant noise with luscious harmonies.
Women who wear their makeup with the pathos
of peasant art settle their cardigans.
Their throats are hoarse, bare light bulbs sting their eyes.
They sing in hot, stale air, lapping the bathos
of a sodden hymn. Old adjectives, old rhymes
fondle their tongues. Amen. The cripple climbs
sweating after her crutch. The pianist fans

a shower of tiny insects from the keys.
The miracle of tongues has been withdrawn.
Let there be silence then. It shall inherit
the suburbs, from Eve's unblemished belly raise
a better race. Lights out. Time to shut down.
Outside, a superb nocturne: street by street
the city lifts its lamps, as if to greet
heaven with luminous gestures of the spirit.

The Wasps

(To Edwin Tanner)

Take up your brush, beloved artist,
  paint me a devil forced to wear
a dress of lavender crochet, picture
  the broomstick limbs and flaming hair.

It is Sunday: a child's Sunday pictures
  lie scattered in loose-fingered grass.
Lolly-coloured assorted faces
  beam at what God has brought to pass.

Under an orange tree her brother
  waddles and crows, absurdly fat.
No godhead skims those rosy features.
  Who said, *Be thou on earth* to that?

Bright from horizon to horizon
  God's eyespace brims with Sunday prayer.
Somewhere above the blue sits Jesus,
  light streaming from his ginger hair.

You told me once you saw in childhood
  a vitreous floater in the sky,
believed it Jesus, and determined
  he should not enter through your eye.

Your brother (Cain? or Abel?) and a
  goanna in the orchard were
enough of scripture when you hunted
  the sky in vain for Lucifer.

Two children of the Devil's party—
  the years frog-march us place by place
to meet in middle life, still probing
  the ambiguities of space.

The sorcerer's apprentice, loathing
  her rival in his harmless play,
implores the Friend of Little Children
  to take and keep him far away.

His plump hand rests in hers. She leads him
  across forbidden garden beds
to trample on the Jesus-gentle
  flowers, and strew their torn-off heads

along the orchard path. No murmur
  comes from the Sinner's Friend on high.
They reach the shadowy barn and stand there
  roofed from the wrath of God's blue eye.

High on a cross-beam, wasps, entombing
  a lightless banquet for their young,
flourish their tiger-stripes to menace
  a child armed with a rake, not stung

before, so fearless. Paint the frantic
  confusion as she strikes at them;
paint the long lavender thread, unravelled
  in headlong flight, from her ripped hem;

paint her supporting blameless Abel
  through Eden spoilt; the mother-saint
forgiving the torn dress, applying
  the bluebag. Pain you cannot paint—

not then, not now, when pain assails you
  in matchless colours, burning bright,
erasing form, dissolving substance.
  You are the shade of its pure light.

When asked once, at an exhibition,
  'Do you believe in God?', for fun
you asked the questioner politely
  'The Blue God, or the Ginger One?'

When I heard this, the colours took me
  through half a lifetime to that day
when I believed the wasps had punished
  my sins in God's mysterious way,

probed my black mocking heart, and taught me
  the foolishness of unbelief.
Now it is pain that I believe in,
  saying at your side, like the Good Thief,

'Remember me.' Remember talking
  late at night of the language game.
'The image of pain is not a picture,'
  said Wittgenstein, 'is not the same

as anything we call a picture.'
  Tell me, who can talk sense with pain
when it becomes the body's language,
  and its unpictured signs remain

a personal mystery affirming
  the person in whose breast reside
all deities, beyond games or language,
  the problem no one gets outside

to solve. With colours in solution
  fix me the opalescent light
of our lost years in solid pigment,
  build me art's heaven in hell's despite:

two children bathed in bluebag, sharing
  the softness of one earthly breast;
old Nobodaddy's social workers
  smoked without mercy from their nest;

picture the tiger-sun declining
  as memory lights with equal flame
vestiges of the pain that leads us
  to truth beyond the language game.

From *Class of 1927*

Religious Instruction

The clergy came in once a week for Religious Instruction.
Divided by faith, not age, we were bidden to be,
(except for the Micks and a Jew) by some curious deduction,
Presbyterian, Methodist, Baptist or C of E.

The Micks were allowed to be useful, to tidy the playground.
But Micah, invited to join them, told Sir 'They'd only
give me a hiding', and stayed inside; moved round
as he chose with his book of Hebrew letters, a lonely

example, among the tender lambs of Jesus,
of good behaviour. Handsome as a dark angel
he studied while the big boys laboured to tease us
with hair-tweak, nib-prick, Chinese burns, as the well

of boredom overflowed in games of noughts
and crosses, spitballs, and drawings so obscene
if Sir had found them they'd have earned us six cuts.
'You give me real insight into original sin,'

said one minister in despair, intercepting some verses
describing him as Old Swivelneck. Beaked like a sparrowhawk
he clawed at his collar and singled me out from his class.
I feared his anger rightly, feared he would talk

to our headmaster, or Sir. I stood in disgrace.
Then a quiet voice from the back interrupted his wrath:
'I am ready to forgive.' 'Who said that? Stand up in your place!'
Micah stood. 'It was said by the Lord God of Sabaoth.'

Then we heard the monitor's footsteps. Saved by the bell!
In a tumult of voices we spilled into sunlight to play,
a host of rejoicing sinners, too young to feel
original darkness under that burning day.

Resurrection

Night-coolness on glittering pasture
still, as my grandmother came
from her watch by a dying neighbour.

She sleepless, and I with my dreams
wide open to quickening day
as I fidgeted under the hairbrush:

'Her soul has taken flight.'
I thought of a humanoid bird
flapping off above sounds of daybreak,

Willie Wagtail, poultry-bicker,
axe-echo, rattle of trace-chains,
to perch on God's golden throne.

—Are we doing some fowls today?
'No, today's not a day for killing,
my dear. Go and catch the butcher.'

So I waved to his cart from the gate
where an elderly rooster, reprieved,
crowed up his second-last morning.

The pick of the cart! He cut
our joint; heavy carcasses swung
from the roof as he trimmed and grunted,

'So the dippy old maid has kicked it!'
My grandmother frowned rebuke.
'Our *neighbour* has *gone to heaven.*'

With the rudeness permitted to one
whose wife was flighty, he barked,
'Don't give me the resurrection!

'Those millions! It stands to reason
there wouldn't be space to hold them.
There simply wouldn't be room!'

I saw pig-pale bodies depending
from hooks in their jaws, and God
with a knife in bilious light

distracted: 'No room! No room!'
But my grandmother, counting her change,
said, 'Remember God isn't a butcher,

'as time will tell.'
I am older
than she was then. Time has told me
less than I need to know.

If eternal life can be given
let me have it again as a child
among those whom I did not love

enough, while they lived. Let me wake,
if I wake at all, on the threshold
of day in my father's house

believing all riddles have answers
night gone   my grandmother walking
head bowed   over glittering pasture.

Night Thoughts

'Hell is for those who doubt that hell exists.'
One of the elohim, with whom I fight
from 4 a.m. to cockcrow, told me this.
He hit me in the thigh for emphasis.
Is it a dream? If so, the dream persists.

I meet him always at the edge of night.
He knows me, but he'll never give his name.
Why should you know, he says. I have to guess
whether he comes to punish or to bless.
I thought once he was death, but at first light

he goes, and I get up. Things are the same
as usual. The sounds of day begin:
the kettle and the news. So it's not death
who comes in the small hours to cramp my breath.
Sleep is extinguished like a candle flame.

Longing for peace, I wrestle, try to pin
my adversary down. Tell me your name.
Tell me, did language lapse when mankind fell?
Tell me, is 'He descended into hell'
a metaphor? The literal truth? Where in

this universe could hell be? He persists:
'Hell is for those who doubt that hell exists.'

Kris Hemensley

the white daisies

(for Ricky Rogers)

the white daisies
are everywhere
& the foxgloves here & there
really do defend the faith

the shadows naturally fall
beside the legion of the apparent
& only night consumes the beautiful—
but as the parent
of divine light

A. D. Hope

Easter Hymn

Make no mistake; there will be no forgiveness;
No voice can harm you and no hand will save;
Fenced by the magic of deliberate darkness
You walk on the sharp edges of the wave;

Trouble with soul again the putrefaction
Where Lazarus three days rotten lies content.
Your human tears will be the seed of faction,
Murder the sequel to your sacrament.

The City of God is built like other cities:
Judas negotiates the loans you float;
You will meet Caiaphas upon committees;
You will be glad of Pilate's casting vote.

Your truest lovers still the foolish virgins,
Your heart will sicken at the marriage feasts
Knowing they watch you from the darkened gardens
Being polite to your official guests.

An Epistle from Holofernes

'Great Holofernes, Captain of the Host,
To Judith: Greeting! And, because his ghost
Neither forgets nor sleeps, peace to her heart!
He, being dead, would play a nobler part;
Yet, being a spirit unpacified, must seek
Vengeance. Take warning then; for souls that speak
Truth to the living, must be fed with blood.
Do not neglect his rites: give him that food
Without which ghosts are powerless to control
Malice which breeds by nature in the soul.
Take down the shining scimitar again;
Slay him a cock, a kid to ease his pain.
For otherwise his talk is double talk,
And he must haunt you. Then, where'er you walk,
Hear his blood dripping from your bag of meat
And, at your table, sitting in your seat,
See the Great Captain's carcass; in your bed
Always upon your pillow grins the Head;

And bloodier whispers that infect the mind
Revenge in dreams the unacted deed of kind.
Think not the Jews nor the Jews' god shall save:
Charms are not sovereign beyond the grave,
And he who warns you, though he wish you well,
Has arms to take and hold you even in Hell.'
Thus in a fable once I spoke to you;
Now other times require I speak it new.
How easy it would be if this were all,
Dear, then the house might totter: it should not fall;
Then, we should utter with our living breath
A healing language from the mouths of death;
In Judith you, in Holofernes I
Might know our legend. But, in days gone by,
This would have been a magic rod whose blow
Broke the parched rocks and made their waters flow.
We should have certainty to conjure with,
Acting the saving ritual of our myth;
The earthquake over, the air sweet and still,
Take courage against this sickness of the will;
For when in former times the myths were true,
For every trouble there was a thing to do:
He, who in faith assumed his Hero's part,
Performed a solemn cleansing of the heart:
The lustral waters, spilling from the bowl,
Poured on the guilty hands and purged the soul;
And sacred dances acted as a spell
To set a lid upon the Hideous Well.
Myths formed the rituals by which ancient men
Groped towards the dayspring and were born again.
Now, though the myths still serve us in our need,
From fear and from desire we are not freed;
Nor can the helpless torment in the breast
Act out is own damnation and have rest.
Yet myth has other uses: it confirms
The heart's conjectures and approves its terms
Against the servile speech of compromise,
Habit which blinds, custom which overlies
And masks us from ourselves—the myths define
Our figure and motion in the Great Design,
Cancel the accidents of name and place,
Set the fact naked against naked space,
And speak to us the truth of what we are.
As overhead the frame of star and star
Still sets rejoicing on the midnight air

Orion's girdle, Berenice's hair,
So when we take our legend for a guide
The firmament of vision opens wide.
Against the sweep of dark and silence lie
Our constellations spread upon the sky.
Plain is the language of those glorious ones;
The meteors flash through their glittering bones;
Freed from the sun of custom, they describe
What, by the daylight vision of the tribe,
We felt, unseeing. We in the mythic night
Know our own motion, burn with our own light,
Study high calm and shining, scorn the more
The beast that winks and snuffles at the door.
Yet the myths will not fit us ready made.
It is the meaning of the poet's trade
To re-create the fables and revive
In men the energies by which they live,
To reap the ancient harvests, plant again
And gather in the visionary grain,
And to transform the same unchanging seed
Into the gospel-bread on which they feed.
But they who trust the fables over much
Lose the real world, plain sight and common touch,
And, in their mythopoeic fetters bound,
Stand to be damned upon infernal ground,
Finding, no matter to what creed they look,
Half their salvation was not in the book.
Then books turn vampires and they drink our blood,
They who feed vampires join the vampire's brood
And, changed to hideous academic birds,
Eat living flesh and vomit it as words.
Our wills must re-imagine what they act
And in ourselves find what the fable lacked.
The myths indeed the Logos may impart,
But *verbum caro factum* is our part.
Thus, though our legend with its proving flame
Burns all to essence, shows in you the same
Temper of ancient virtue, force of will,
That saved the trembling people on the hill;
Though I myself in Holofernes know
Your bloody and greedy and insensate foe
And, at my feast, hear a relentless voice
Declare my grim dichotomy of choice:
Sound a retreat, or move to one event:
The headless carrion rolling in the tent;

Yet imaged new the fable is not plain:
Though Judith live and though the foe be slain,
Ours is a warfare of a different kind
Pitched in the unknown landscape of the mind,
Where both sides lose, yet both sides claim the day
And who besieges whom is hard to say;
Where each, by other foes encircled round,
Hears in the night far off the bubbling sound
Of the sweet springs that are to both denied,
And sees false watch-fires crown the mountain-side.
Where shall we turn? What issue can there be?
Through the waste woods we searched and found the Tree,
Sole of its kind, bowed with its precious fruit;
And lo, the great snake coiling round the root!
Was all our toil, our patience, then, for this?
Our prayers translated to a brutal hiss?
Our desperate hopes, the fears and dangers passed,
To end in death and terror at the last?
Reach me your hand; the darkness, gathering in,
Shrouds us—for now the mysteries begin:
The world we lost grows dim and yet on high
Figures of courage glitter in the sky;
And, though a desert compass us around,
Layers of water lie beneath that ground;
The fissure in the rock that sets them free
Feeds and refreshes our Forbidden Tree.
Already, though we do not feel it yet,
The unexpected miracle is complete;
Already, through the midnight hours, unseen
They rise and make these barren places green
Till the parched land in which we lost our way
Gives grace and power and meaning to the day,
Renews the heart, gives joy to every act
And turns the fables into living fact.
If in heroic couplets, then, I seem
To cut the ground from an heroic theme,
It is not that I mock at love, or you,
But, living two lives, know both of them are true.
There's a hard thing, and yet it must be done,
Which is: to see and live them both as one.
The daylight vision is stronger to compel,
But leaves us in the ignorance of hell;
And they, who live by starlight all the time,
Helpless and dangerous, blunder into crime;
And we must learn and live, as yet we may,
Vision that keeps the night and saves the day.

A Bidding Grace

For what we are about to hear, Lord, Lord,
The dreadful judgement, the unguessed reprieve,
The brief, the battering, the jubilant chord
Of trumpets quickening this guilty dust,
Which still would hide from what it shall receive,
Lord, make us thankful to be what we must.

For what we are now about to lose, reprove,
Assuage or comfort, Lord, this greedy flesh,
Still grieving, still rebellious, still in love,
Still prodigal of treasure still unspent.
Teach the blood weaving through its intricate mesh
The sigh, the solace, the silence of consent.

For what we are about to learn too late, too late
To save, though we repent with tears of blood:
The innocent ruined, the gentle taught to hate,
The love we made a means to its despair—
For all we have done or did not when we could,
Redouble on us the evil these must bear.

For what we are about to say, urge, plead,
The specious argument, the lame excuse,
Prompt our contempt. When these archangels read
Our trivial balance, lest the shabby bill
Tempt to that abjectness which begs or sues,
Leave us one noble impulse: to be still.

For what we are about to act, the lust, the lie
That works unbidden, even now restrain
This reckless heart. Though doomed indeed to die,
Grant that we may, still trembling at the bar
Of Justice in the thud of fiery rain,
Acknowledge at last the truth of what we are.

In all we are about to receive, last, last,
Lord, help us bear our part with all men born
And, after judgement given and sentence passed,
Even at this uttermost, measured in thy gaze,
Though in thy mercy, for the rest to mourn,
Though in thy wrath we stand, to stand and praise.

From *A Letter from Rome*

Unlike that desultory scenic stroll
Which robs his earlier cantos of their force,
This moves, with sure direction and control,
In towards the centre, back towards the source.
Its theme is destiny and Rome its goal;
And yet it does not stop with Rome; the course
Of history retraced, it moves at last
Into the savage, pre-historic past.

It ends with Nemi and the Golden Bough.
What instinct led him there? I like to think
What drew him then is what has drawn me now
To stand in time upon that timeless brink,
To sense there the renewal of a vow,
The mending of a lost primordial link.
These may be only fancies, yet I swear
I felt the presence of the numen there.

There's nothing now at Nemi to evoke
Sir James G. Frazer's memorable scene:
The sleepless victim-King, the sacred oak;
A market garden spreads its tidy green
Where stood Diana's grove; no voices spoke;
There were no omens; cloudless and serene
The sun beat harshly on the drowsing lake;
And yet I felt my senses wide awake,

Alert, expectant—as we scrambled down
The crater from the village to the shore
And strolled along its path, we were alone;
And, picnicking among the rocks, I saw
No cause for these sensations. Yet I own
A tension grew upon me more and more.
What Byron felt as calm and cherished hate
For me was more like force, insistence, fate.

And under this impulsion from the place,
I seemed constrained, before I came to drink,
To pour some wine upon the water's face,
Later, to strip and wade out from the brink.
Was it a plea for chrism or for grace?
An expiation? More than these, I think
I was possessed, and what possessed me there
Was Europe's oldest ritual of prayer.

But prayer to whom, for what? The Intervention
Did not reveal itself or what it meant.
The body simply prays without 'intention',
The mind by the bare force of its assent.
That 'higher, more extended comprehension',
Which Byron, writing after the event,
Felt necessary to explain brute fact,
Came by mere power of my consenting act.

Well, let it pass: I have no views about it;
Only I sensed some final frontier passed,
Some seed, long dormant, which has stirred and sprouted,
Some link of understanding joined at last.
I may have been deluded, but I doubt it
Though where the series leads I can't forecast.
Laugh at these intimations if you will;
The days go by and they are with me still.

Faustus

Laying the pen aside, when he had signed,
'I might repent, might yet find grace,' he said,
'What could you do?' The Devil shook his head,
'You're not the first, my friend: we know your kind.

'Logic, not justice, in this case prevails:
This bond can't be enforced in any court.
You might prove false as hell, but have you thought
The fraud may damn you, though the promise fails?'

'Suppose I use these powers, as well I may,'
Said Faustus then, 'to serve the cause of good!
Should Christ at last redeem me with his blood
You must admit there'd be the devil to pay.'

The Devil laughed and conjured from the air
A feast, a fortune and a naked bed.
'Suppose you find these powers use you instead!
But pun your way to heaven, for all I care.

'We could have had your soul without this fuss.
You could have used your wits and saved your breath,
Do what you like, but we at least keep faith.
You cheated God, of course; you won't cheat us.'

Faustus unclasped the Book: when that first hour
Struck on his heart, a fragment broke away.
What odds? With four and twenty years to pay
And every wish of man within his power!

He asked to know: before the words were said
Riddles that baffled Kepler all lay bare;
For wealth, an argosy walled in his chair;
For love and there lay Helen in his bed.

Years passed in these enchantments. Yet, in fact
He wondered sometimes at so little done,
So few of all his projects even begun.
He did not note his will, his power to act

Wither, since a mere wish would serve as well,
His reason atrophy from day to day
Unexercised by problems, Love decay
Untried by passion, desire itself grow stale,

Till he, who bought the power to command
The whole world and all wisdom, sank to be
A petty conjurer in a princeling's fee
Juggling with spells he did not understand

And when, at last, his last year came, and shrank
To a bare month and dwindled to an hour,
Faustus sat shuddering in his lamp-lit tower
Telling the time by seconds till time went blank.

Midnight had come: the fiend did not appear;
And still he waited. When the dawn began
Scarce crediting his luck he rose and ran
And reached the street. The Devil met him there.

It was too much. His knees gave way. He fell.
'The bond? . . . My soul?' Quite affable the fiend
Helped him to rise: 'Don't fret yourself my friend;
We have your soul already, quite safe, in Hell.

'Hell is more up-to-date than men suppose.
Reorganized on the hire-purchase plan,
We take souls by instalment now and can
Thus save the fuss and bother to foreclose.

'And since our customers prefer, you know,
Amortized interest, at these higher rates,
Most debts are paid in full before their dates.
We took your final payment months ago.

'But, as I say, why fret? You've had your fun.
You're no worse off without a soul you'll find
Than the majority of human kind,
Better adjusted, too, in the long run.'

Back in his tower Faustus found all bare.
Nothing was left. He called: the walls were dumb,
Drawing his knife, he stalked from room to room
And in the last he found her, waiting there,

That fabulous Helen his magic art had won.
Riches and power, she was their sum and prize;
Ten thousand years of knowledge were in her eyes
As first he cut her throat and then his own.

The Mystic Marriage of St Catherine of Alexandria

While Jesus was still a baby she fell in love,
Though not yet born when he was crucified;
She waited four centuries more to be his bride.
Those legends, in fact, will not bear thinking of,
Where not even fiction with faith goes hand in glove.

Yet there was a truth in all that pious confusion,
Waiting like seed in desert sands for rain;
Though time was a tongueless tocsin tolled in vain,
Her virgin visions were not all delusion,
But causes pointing to no foregone conclusion.

This is certain at least: at her conversion she saw
Seven golden candlesticks and in their light,
Muscled like a lion, a man; his hair was white
Like wool; his eyes like flame; his jaw
Clamped like a sheath upon the sword of law.

Yes, he was terrible, tall, ruddy and young;
The meeting of their eyes was night with noon.
His glance said: 'Wait for me!' and hers asked: 'Soon?'
As the meeting of many waters then his tongue
Cried: 'No, you must wait till this world's knell is rung.'

And when they baptised her, as on breast and face
The holy, eclipsing water flashed and fell,
The redemptive drops enchanted her as well,
Dissolved the Bridegroom's male and menacing grace
And Mary his mother sat smiling in his place.

A naked infant, laughing on her knee,
Leaned, touched her hand; the baby fist took hold
And loosed, and there she felt a ring of gold
Clasped on her finger, not to be pulled free,
While her flesh shivered in mortal ecstasy.

A needle of cureless love transfixed her heart
And the young girl, so like her, held the child
Towards her, still crowing and gurgling; and she smiled
Saying: 'See, we have both been chosen and set apart.
Now, from this moment, your martyrdom will start.

'The agony of inscrutable centuries
Of waiting will be your torment, past relief
But He will come at last.' There was a brief
Dazzle and darkness; when she opened her eyes
She found the brethren staring with surprise.

The priest said: 'While you were out of the body, this hour,
There was a voice as though from heaven: We heard:
"She is blessed among women. To wed the Word
Shall be her lot; an immarcescible flower
Her elective sign; three visions her bridal dower."'

She made no answer; a glory, it is said,
Shone from her face, her hands, her naked feet.
The brethren and strangers peering from the street
Gaped as at Lazarus risen from the dead.
She knelt and took the blessed wine and bread.

The legends tell next of incredible things:
Of fifty pagan philosophers rendered dumb;
Her shattering of the wheel of martyrdom;
Her elegant frustration of the King's
Crazed lust; her perlustration on the wings

Of forty angels to Sinai—what is true
Is this: she outlived the people of her time.
A perpetual virgin lies within reason and rhyme;
But she was a perpetual beauty too,
Seventeen in all its bloom, its dazzle, its dew.

She walked the great city daily; a golden haze
About her; no shoulder jostled her in the crowd;
Wolfish men answered her gently and the proud
Gave what she asked. Her inward-centred gaze
Fixed the two visions that filled her nights and days.

Yet no one seemed to notice her; as though
Closed in a cloud she wandered among men
The centuries came and went. Rome fell; and when
Islam arose and Egypt felt the blow
She fled to Byzantium. No one saw her go.

There, age after age, she toiled at the immense
Task of unflagging patience she called love.
The domes of the Holy Wisdom, poised above
Her praying posture, figured the suspense
Of passion on the long agony of sense.

An agony which at last she could not fight
And prayed for death, for peace: 'Have pity, Lord!'
But he: 'I bring not peace, love, but a sword;
I shall come in terror, in glory and in might;
Yet you shall be with me in Paradise this night.'

It was the year the Turk stood at the gate;
The day Constantinople drowned in blood.
Going to her lodging from the church she trod
All the way on corpses, for the slaughter was great;
But, treading on death, sang her magnificat:

'In my Redeemer now am I glorified!
Now the doves in the clefts of the rock prepare their nests.
Let him come: he shall lie all night between my breasts.
Like an army with banners, the morning stars in their pride
The Bridegroom approaches at last to claim his bride!'

She climbed the twelve steps to her garret room;
Took off her clothes; combed out her radiant hair;
And, breaking a box of very ancient myrrh,
Anointed her body for bridal and the tomb
Crying: 'Eli, Eli! Come now, Lord Jesus, come!'

He said: 'I am here, behold, my love, my bride!'
And, terrible in his naked might, he came;
Gathered her harvest in his devouring flame;
Then saying: 'It is finished!' gave up the ghost and died.
Wondering, she touched his wounded hands, his side.

So the third vision left her. It is said
In the burning city at dawn, or just before,
A janizary for loot broke down her door;
Found an old crone asleep upon the bed
And, having searched the room, cut off her head.

Visitant

Earth swings away to the cold.
Though I have what I came here to find,
Time changes and alters the mould.
As a new age replaces the old
I feel the world leave me behind.

It is not my world any more;
But of course was it ever mine?
Bred up to a different law,
I came from a distant shore
To watch, to appraise, to divine.

Yet much that I saw became dear;
Some few were close to my heart;
Although it was perfectly clear
I was a stranger here
Standing aloof and apart.

Now it is time to return,
I shall miss this world more than I thought.
All I came merely to learn
Holds me now with such love and concern,
To whom do I make my report?

Kate Jennings

Saint Munditia

Saint Munditia (the patron saint
of lonely women) is a relic bound
to inspire affection. Her bones
are barnacled with gold and jewels.
Her skull is stuck in a perpetual grimace.
Cat-yellow eyes start out of their sockets.
And in her right hand, fixed at the ready, a writing quill.
Her elaborate hideousness made me laugh,
and, in my mind, her image prospered.
Sometimes I petition her.

Saint Munditia,
they broke your legs to fit you in your showcase,
and I am broken, too.
There's nobody to whom either of us can address a letter.
My humour is as brittle as you.

Saint Munditia,
why don't we sell your jewels
and go see the world, live it up big?
We would make a striking couple.
People are sure to befriend us
for our money and our enthusiasm for life.

Saint Munditia,
damn this loneliness. Damn my seedy guilt,
my laziness, my fears. Damn my cracked mind.
Tell me why it should be so.
Saint Munditia, comfort me.

Evan Jones

Servetus: October 27th, 1553

Grey morning, clear October day,
As from the cell of four long weeks
Of doubt, privation, fear of death,
They dragged the living, dying husk.
And where the crowds defined his way
The last leaves were received by earth.

So through the city streets he passed,
Weak with his fear and fever; but
Thrown down before the City Hall
And asked if he would not retract,
Held firm, addressed himself to Christ;
And in the fire at Champel.

Poor mortal, martyred in the fall
Of seasons, cities, centuries,
How like your killers, unaware
Of this event as history:
Each man's own ministry, and all
That protest, blazing in this fire.

And Calvin, in his narrow room,
Against his scorned, spurned body's pain
Cast all his faculties on God
(Though still presiding on that scene)
And rendered final praise to Him;
Then turned to work too long deferred.

Antigone Kefala

Alter Ego

The cardinal was waiting at the corner for a bus.
Dusk, empty Sunday streets,
the Kings' Hotel embossed in phosphorescent yellow.

He seemed so small without his gear,
the ritual gestures and the red silk gowns
now preserved in moth balls.
The brilliant upper crust that formed the man,
suave, full of a worldly vision that had learnt
to by-pass suffering, to offer eternity
as if he had the piece at home,
and cut it into measures that would suit him.

Worship

They ate the soil, slept with it,
the scent coursing their blood
till they were filled with earth,
took the sheep for lovers
prayed in mutilated voices,
in harsh goat tongues they sang
of rocky sites, hard winters
echoes of wind at night
in the rough trees.

The arms of the white statue
on the peak
stretching like giant horns
in the spring sun.

Henry Kendall

Dedication

To a Mountain

To thee, O Father of the stately peaks,
Above me in the loftier light—to thee,
Imperial brother of those awful hills
Whose feet are set in splendid spheres of flame,
Whose heads are where the gods are, and whose sides
Of strength are belted round with all the zones
Of all the world, I dedicate these songs.
And, if within the compass of this book,
There lives and glows *one* verse in which there beats
The pulse of wind and torrent—if *one* line
Is here that like a running water sounds,
And seems an echo from the lands of leaf,
Be sure that line is thine. Here in this home
Away from men and books and all the schools,
I take thee for my Teacher. In thy voice
Of deathless majesty, I, kneeling, hear
God's grand authentic gospel! Year by year,
The great sublime cantata of thy storm
Strikes through my spirit—fills it with a life
Of startling beauty! Thou my Bible art
With holy leaves of rock, and flower, and tree,
And moss, and shining runnel. From each page
That helps to make thy awful Volume, I
Have learned a noble lesson. In the psalm
Of thy grave winds, and in the liturgy
Of singing waters, lo! my soul has heard
The higher worship; and from thee indeed
The broad foundations of a finer hope
Were gathered in; and thou hast lifted up
The blind horizon for a larger faith!
Moreover, walking in exalted woods
Of naked glory—in the green and gold
Of forest sunshine—I have paused like one
With all the life transfigured; and a flood
Of light ineffable has made me feel
As felt the grand old prophets caught away
By flames of inspiration; but the words
Sufficient for the story of my Dream

Are far too splendid for poor human lips!
But thou to whom I turn with reverent eyes—
O stately Father whose majestic face
Shines far above the zone of wind and cloud
Where high dominion of the morning is—
Thou hast the Song complete of which my songs
Are pallid adumbrations! Certain sounds
Of strong authentic sorrow in this book
May have the sob of upland torrents—these,
And only these, may touch the great World's heart;
For lo! they are the issues of that Grief
Which makes a man more human, and his life
More like that frank exalted life of thine
But in these pages there are other tones
In which thy large superior voice is not—
Through which no beauty that resembles thine
Has ever shone. *These* are the broken words
Of blind occasions when the World has come
Between me and my Dream. No song is here
Of mighty compass; for my singing robes
I've worn in stolen moments. All my days
Have been the days of a laborious life;
And ever on my struggling soul has burned
The fierce heat of this hurried sphere. But thou
To whose fair majesty I dedicate
My book of rhymes—thou hast the perfect rest
Which makes the heaven of the highest gods!
To thee the noises of this violent time
Are far faint whispers; and, from age to age,
Within the world and yet apart from it
Thou standest! Round thy lordly capes the sea
Rolls on with a superb indifference
For ever: in thy deep green gracious glens
The silver fountains sing for ever. Far
Above dim ghosts of waters in the caves,
The royal robe of morning on thy head
Abides for ever! evermore the wind
Is thy august companion; and thy peers
Are cloud, and thunder, and the face sublime
Of blue midheaven! On thy awful brow
Is Deity; and in that voice of thine
There is the great imperial utterance
Of God for ever; and thy feet are set
Where evermore, through all the days and years,
There rolls the grand hymn of the deathless wave.

Peter Kocan

Cathedral Service

I'm only here because I wandered in
Not knowing that a service would begin,
And had to slide into the nearest pew,
Pretending it was what I'd meant to do.

The tall candles cast their frail light
Upon the priest, the choir clad in white,
The carved and polished and embroidered scene.
The congregation numbers seventeen.

And awkwardly I follow as I'm led
To kneel or stand or sing or bow my head.
Though these specific rites are strange to me,
I know their larger meaning perfectly—

The heritage of twenty centuries
Is symbolised in rituals like these,
In special modes of beauty and of grace
Enacted in a certain kind of place.

This faith, although I lack it, is my own,
Inherent to the marrow of the bone.
To this even the unbelieving mind
Submits its unbelief to be defined.

Perhaps the meagre congregation shows
How all of that is drawing to a close,
And remnants only come here to entreat
These dying flickers of the obsolete.

Yet when did this religion ever rest
On weight of numbers as the final test?
Its founder said that it was all the same
When two or three were gathered in his name.

Anthony Lawrence

God is with me as I write this down

God is the poetry caught in any religion,
caught, not imprisoned.
Les Murray

God is with me as I write this down,
though he would not look over my shoulder
or ask me to read what I've done.

Yesterday he was with me at the river
as I watched the redwinged fisher take
its unprotesting catch from the water.

Between the willow's beaded veil
and the river's grey light, God moved
as the day moved me, speaking outside time

through the poetry of each day's importance,
leaving a prayer to smoulder on my tongue,
kneeling with me until my prayer had gone.

God is with me as I write this down,
though he would not ruffle my hair with visions
or enter where the muse reclines.

He takes me laughing from the seriousness of my work:
the black lines arranged with love on the page;
the black heart trembling with release and rage;

my poetry my love.

Geoffrey Lehmann

From *Spring Forest*

Mother Church

I came into this century a Catholic
and shall leave it with no belief,
like a hundred million others.
Faith like mineral salts
is leaching out of the soil.
A Pope was afraid to speak out
against blackshirts and murderers
and did nothing
as the world drifted into war.
A later Pope, lost in his own dogma,
was too craven to speak the word
that would release the unconceived
from a birth without future.

Transubstantiation, free will—
mouthfuls of nothing.
Mother Church
when our earth cried out
you had nothing to say.

Where is your 'life after death'?
What about the life now?
'Infallible'—1869.
How silly it all sounds to our ears.
I can only laugh and cry.

Yet if outsiders attack you
I'll strike them down.

Witnesses

Jehovah's Witnesses, you say with a look of pity.
Well, I'm not one,
but they may well look with pity
at you.

While you and I
despair for the human race
(numbed as reported horrors bombard us)
they can say:
'It is written, this world is mad,
we are not surprised
as we see these ghosts
chasing each other with cutlasses across the quicksands,
the nations sinking.'

They refuse to take oaths—
no man can tell the truth.
Bellboys, plumbers, bus conductors,
they live simply in a time of madness,
accumulate few assets
and wait for the day of judgment.

So many of us die trying to right the world,
the widow in her apartment
shrill with anger at students,
the old socialist cooking toadstool soup for 'The Bosses'.

These Witnesses for all their crazy door-knocking
and Old Testament readings
proclaim one truth—
we are witnesses of the conflagration,
the fires are happening already, all around us.
Our possessions and protests are useless,
our despair is useless.

I am walking down from my father's hill
in another direction
among clean tussocks and granite,
free in the Antarctic night.

James McAuley

To the Holy Spirit

Leaving your fragrant rest on the summit of morning calm,
Descend, Bird of Paradise, from the high mountain;
And, plumed with glowing iris along each curving wire,
Visit in time our regions of eucalypt and palm.

Dance, prophetic bird, in rippling spectrums of fire,
Ray forth your incandescent ritual like a fountain;
Let your drab earthly mate that watches in morning calm
Unseen, be filled with the nuptial splendours of your desire.

Engender upon our souls your sacred rhythm: inspire
The trembling breath of the flute, the exultant cosmic psalm,
The dance that breaks into flower beneath the storm-voiced
 mountain;
Array in your dazzling intricate plumage the swaying choir.

New Guinea

(In memory of Archbishop Alain de Boismenu, MSC)

Bird-shaped island, with secretive bird-voices,
Land of apocalypse, where the earth dances,
The mountains speak, the doors of the spirit open,
And men are shaken by obscure trances.

The forest-odours, insects, clouds and fountains
Are like the figures of my inmost dream,
Vibrant with untellable recognition;
A wordless revelation is their theme.

The stranger is engulfed in those high valleys,
Where mists of morning linger like the breath
Of Wisdom moving on our specular darkness.
Regions of prayer, of solitude, and of death!

Life holds its shape in the modes of dance and music,
The hands of craftsmen trace its patternings;
But stains of blood, and evil spirits, lurk
Like cockroaches in the interstices of things.

We in that land begin our rule in courage,
The seal of peace gives warrant to intrusion;
But then our grin of emptiness breaks the skin,
Formless dishonour spreads it proud confusion.

Whence that deep longing for an exorcizer,
For Christ descending as a thaumaturge
Into his saints, as formerly in the desert,
Warring with demons on the outer verge.

Only by this can life become authentic,
Configured henceforth in eternal mode:
Splendour, simplicity, joy—such as were seen
In one who now rests by his mountain road.

Pietà

A year ago you came
Early into the light.
You lived a day and night,
Then died; no-one to blame.

Once only, with one hand,
Your mother in farewell
Touched you. I cannot tell,
I cannot understand

A thing so dark and deep,
So physical a loss:
One touch, and that was all

She had of you to keep.
Clean wounds, but terrible,
Are those made with the Cross.

Father, Mother, Son

From the domed head the defeated eyes peer out,
Furtive with unsaid things of a lifetime, that now
Cannot be said by that stiff half-stricken mouth
Whose words come hoarse and slurred, though the mind is sound.

To have to be washed, and fed by hand, and turned
This way and that way by the cheerful nurses,
Who joke, and are sorry for him, and tired of him:
All that is not the worst paralysis.

For fifty years this one thread—he has held
One gold thread of the vesture: he has said
Hail, holy Queen, slightly wrong, each night in secret.
But his wife, and now a lifetime, stand between:

She guards him from his peace. Her love asks only
That in the end he must not seem to disown
Their terms of plighted troth. So he will make
For ever the same choice that he has made—

Unless that gold thread hold, invisibly.
I stand at the bed's foot, helpless like him;
Thinking of legendary Seth who made
A journey back to Paradise, to gain

The oil of mercy for his dying father.
But here three people smile, and, locked apart,
Prove by relatedness that cannot touch
Our sad geometry of family love.

One Tuesday in Summer

That sultry afternoon the world went strange.
Under a violet and leaden bruise
The air was filled with sinister yellow light;
Trees, houses, grass took on unnatural hues.

Thunder rolled near. The intensity grew and grew
Like doom itself with lightnings on its face.
And Mr Pitt, the grocer's order-man,
Who made his call on Tuesdays at our place,

Said to my mother, looking at the sky,
'You'd think the ending of the world had come.'
A leathern little man, with bicycle-clips
Around his ankles, doing our weekly sum,

He too looked strange in that uncanny light;
As in the Bible ordinary men
Turn out to be angelic messengers,
Pronouncing the Lord's judgments why and when.

I watched the scurry of the small black ants
That sensed the storm. What Mr Pitt had said
I didn't quite believe, or disbelieve;
But still the words had got into my head,

For nothing less seemed worthy of the scene.
The darkening imminence hung on and on,
Till suddenly, with lightning-stroke and rain,
Apocalypse exploded, and was gone.

By nightfall things had their familiar look.
But I had seen the world stand in dismay
Under the aspect of another meaning
That rain or time would hardly wash away.

Music Late at Night

Black gashes in white bark. The gate
Is clouded with spicy prunus flowers.
The moon sails cold through the small hours.
The helpless heart says, hold and wait.

Wait. The lighted empty street
Waits for the start of a new day,
When cars move, dogs and children play.
But now the rigid silence is complete.

Again that soundless music: a taut string,
Burdened unbearably with grief
That smiles acceptance of despair,

Throbs on the very threshold of spring
In the burst flower, the folded leaf:
Puzzling poor flesh to live and care.

Hugh McCrae

Down the Dim Years

Clean running wave and sunward soaring flower,
  The great hot sky, the colours of the wood
Troubled with shadow, and the sudden shower
  Of heavenly fire across the solitude
Of some green praying-place with steps of stone
  As sorrowfully white as winter moon.
The thin far calling of a trumpet blown
  To dead men hunting in the afternoon.

The pink-faced boy who kissed the crying witch
  Upon her wet flat mouth, and, wailing, went
Down into hell . . . while she, a snuffling bitch,
  This way and that, bewrayed the tasselled bent,
Howling the pain of Doom.

                    The quiet Christ
Leading His children surely through the dark
  By the clear lantern of His wounds, to tryst
With God; of men their refuge and their ark.

Nan McDonald

Sunday Evening

The hollows of the sandy track are cold
With lilac shadow now and evening's breath
Under our homeward feet, and to the land's last edge
From the wide miles of bush the sunlight dies,
Gathering its gold in the unclouded west,
As the blue draining from the high, pale sky
Sinks to the woodsmoke mist along the creeks.
This is the hour when all things are drawn in
And made more deep, when in my heart it swells
And floods once more in peace, the old, strong tide
Too sad for ecstasy, too calm for pain,
Too deep to rise because one day was gold
And now is done. It is the gathered light
Of many sunsets, ending other days—
Those days like jewels strung far back through life,
As far as memory, days of air and sun
In fern-green gullies bird- and water-sweet,
On uplands brown above kingfisher seas;
Days drunk with freedom, more than the escape
From streets and walls—fresh spring of youth that flows
Bright from the harsh breast of the ancient bush,
When the blank, loveless face of crowds is changed
For the dear small world of friends, the barriers down;
Days when the sun, grown kindly in the north,
Slants through the shining leaves, soft yet so clear
Far hills curve clean from sky all washed with light
As though a globe of crystal held our joy
Safe from tomorrow and from yesterday.

They do not know, those envied ones who live
Whole years in beauty, how the splendour breaks
Out of our sabbaths and our holydays
Set in grey weeks. Their treasures come so fast
They have no time to polish and adore
Each stone, as we who turn them in our hands
—The peacock sheen, the fireglow at the heart—
Through dull, tired days, and sweetly tell them over
To light long nights of sickness and of fear.

So with no outcry, but with quiet hearts
We watch the sun set on our happiness,
The clear gold chill to primrose-green, low-edged
With smoky orange, and the dark-blue bush
Fold all its secrets in the black of night.
Though that sun rise with a changed and a worse face
And bring our fears to truth, despair, love's end,
New ills more evil than the most evil past,
These it can add, but cannot take away
This day, safe now with all our perfect days.
And though the crooked worm in the brain will say,
'Here's the old tale again, the slave is free,
The dead shall rise—but chain and grave hold yet',
His grinding jaws will try our hoard in vain.
These are no fading flowers, but opals mined
From the blind rockface of man's misery,
The choking dust of man's futility,
Not by our will alone, not by our strength,
But by that power which time the moth and pain
The rust and death the thief shall not defy;
These are the gift of Love, for ever ours,
These shall make up our incorruptible crown.

The Barren Ground

I think of it, that high, bare place in the mountains,
Always under a cool grey blowing sky
Where the eagle hangs black and alone—clean wind whistling
Through tussock and long low swell of scrub, and through
Hollows the rainy centuries have worn
In the huge ruinous heaps of silent stone,
I trace on the map the way from the blue-edged coast
Across green, where the rich farms lie, through the deepening brown
That marks the rising ground, till the clear road fades
And a broken track goes climbing on to that space
Empty of all but the three words standing plain,
And I think I could lie down there, and be at peace.

It was not named for a blue dog, a dead horse,
Nor for some dark tribe's thought that haunts the ear
And still eludes the brain; no bold explorer
Made record of his desperate journey there;
No settler left his name to cry when the bush

Has flowed back across the furrows and the cold hearth,
When the fruit-trees have gone wild and the children are scattered,
'By sorrow and sweat I was conqueror of this earth.'
No, it was not named for glory or possession,
The foolish dreams of man, whose stiffening hand
Lets the bright coin drop and follows it down to dust,
But simply for itself—The Barren Ground.

How lovely its fall on the ear bewildered with words,
On the mind sick with the eddies of thought, for at last
The eyes are filled with the glitter of light on the leaves,
With the million-petalled flowers, and can only thirst—
Turning from life grown rank and smelling of death—
With a great thirst for the truth of barrenness.
For what is life, or what is death, who knows?
But all men know that peace is only peace.

Yet it may not be as I dream—three words on the map
Are all my knowledge, and that I have seen as well
Through a gap in the hills, rising far and strange in the evening,
The darkening ramparts of its eastern wall.
But the heart has its own routes and its journeys' ends,
And sooner or later strikes, though travelling blind,
The rocky track where the long road gives out
And comes on a grey day to its Barren Ground.

The Return

So after twenty years we stand
By the old well, and hear the wind
Sweep over from the bay once more
Slashing the broad banana leaves,
And see again the red ploughed land
Run to the ridge of mango trees
Dark in the blazing light we knew
That summertime of drought and war.

The three tall ringbarked gums are gone
That stood so white on the black sky
When day by day the stormclouds rose
And brought no rain. Again the sun
Cracked the bare ground, the thirsty cows
Stared while we pumped brown water here
And knew the well was nearly dry,
But did not let our thoughts run on.

For day by day, like strokes of doom
Beaten on some great brazen gong,
Out of the north the news came in,
The ships went down, the cities fell,
The towering wave of nightmare rolled
On towards our naked coast, and kin
And friend died in its path, and some
Alive descended into hell.

And yet how much we laughed those days—
Loud above pity, fear, and grief,
Our laughter echoes in this place,
We cannot hush that memory
But ask no pardon of the dead
Who also knew the heart's strange ways
Being young in such a time, when life
Ran like a rip-tide out to sea.

And late, but not too late, the rain
Came in grey columns marching on—
Remember in the dripping night
The full frog-chorus, dumb so long?
The well flowed over, the grass grew
So fast in our rejoicing sight
We brought the cows to milking soon
Soaked to our waists in morning dew.

And not too late, in that last ditch
Of swampy shore and jungle track,
Sick and exhausted, there were men
Who stood, and held the enemy.
We heard his scream of triumph die
As inch by inch they forced him back
Across the barbarous range; his ships
Were broken in the Coral Sea.

The years we never thought to have
We shall not weigh, at this return,
For loss or gain, for waste or joy,
Knowing the seed grows secretly,
The harvest is with God; nor cry
To wake the sleepers in the grave.
If we have one wish, let it be
That this fierce sun may strike and burn

The mould that grows in softer days
In this world's peace that is not peace,
The dream that we are safe from death,
That we have time for pettiness.
Show us again, pierced through with light,
The sacraments of bread and breath,
The miracle of being blaze
On the dark verge of mortal night.

The Last Mile

Let night fall now, we cannot lose our way.
Dusk gathers under crouching trees, and westward
Sun-heat dies in red embers, and the air
Breathes cold as mountain creeks we crossed today,
But our road lies plain ahead, the steep turns gone,
And labouring heart and driven muscles ease
To a steady rhythm, untiring, on and on.

Darkness would have been danger, two miles back,
Hard to the rockface, the sheer drop below,
Or three miles, in that gully drowned in leaves,
Fearing we might have chosen the wrong track
So faint it showed; but we have safely come
In sight of pines and smoke from evening fires.
We could go blindfold now, and still reach home.

And in the untroubled mind all joys return
From the spent hours, and float like sunset clouds:
Heathland of honey-flowers and a bird crying,
Spun mist of waterfalls, white sand and fern,
The enormous solemn shade of afternoon
Filling the valley with dark blue, the gold
Of weathered cliffs, more ancient than the moon.

They stream with more than one day's light, they shine
With a child's wonder, for I came this way
When there were arms to lift me if I tired,
And as a girl, and still they glow like wine,
And youth gone by they were healing and release
From sickness of life marred—so many times
Alike drawn in to the last mile in peace.

God who has showered me with blessings all my years
Now gives me this as well, to walk once more
Recalling sunlight and water, leaf and stone,
And perils past, childhood and youth and tears,
On the old road through the bush at the end of day
With the living and the dead I love beside me,
And if night falls, I cannot miss the way.

Kenneth Mackenzie

How near, o god . . .

How near, o god, the consummation
of my most desperate ambition
to praise you with the purest praise
during the darkest of my days?

How far, o god, the realisation
of my unpardonable defection?
How soon shall I escape the mesh
of weaknesses of mind and flesh?

Let me have time for absolution,
for chaste living and ablution
in body's peace and mind's content
so that I may be continent.

Let me have time for contemplation,
prayer of some sort, and oblation—
squeezing from the grape of strife
words of my love for you—for life.

No body . . .

No body . . . god above!
when mind's all ripe to love,
all it can claim to have
is this engulfing grave,
this buried wooden bed
lined with a skin of lead.

Selfish indeed is death
when, having stilled the breath,
it cannot be content
to know that life is spent
but calls the mind its own
and makes it live alone.

These things I learned one night
an hour before daylight
when having lain long dead
within my cooling bed,
I found that mind lives on
when body's life is gone.

David Malouf

Metamorphoses

The forbidden name: spelled out on a screen as wheatgrain
flicker, scribbled on pavements, the initials
that blood makes trailed from knives. A blond grease-monkey
slides out under axle-bars and wears the Smile,

a deus ex machina. The Word at this point
enters someone's head, a stutter finding
speech in the slow vernacular of locals.
An eye among rosebuds winks from an open fly

and half a county quickens, so many siblings,
tongue-tied, snub-nosed, tufted, gravely conceiving
of father as a hepped-up Maserati.
A buzz some schoolboy swats is left for spiders

to drag out of the sun: announcing *OM*,
his name, too low to cut through conversations.
No words pass between species. Nothing changes.
This angle of light the gospels will get wrong.

J. S. Manifold

From *Red Rosary*

Death of Stalin

North to the reindeer herds, the snowbound dark,
Mammoth-tusk carvings and enormous pines;
South to the great canals, the silk, the vines,
The turbanned heads as brown as wattle-bark;

East where the slant-eyed fishermen embark,
And tigers prowl between the silver-mines;
West to the wheatlands where the roaring lines
Of tractors wipe away the invaders' mark;

Such is his vast memorial's extent!
Here—like a fighter-plane, his petrol spent,
But straining dauntless towards a friendly drome

Whilst all his victories yet blaze in air—
Here at the dawn-lit first perimeter
Of Communism Uncle Joe reached home.

Astronauts

Yours too is an exacting life, you know.
The ship goes spinning on its way thro' space
While calmly you explore its carapace
Or tinker with its innards down below.

Clear in your sight untrodden planets glow
And comets pass—but what if one gave chase?
Say what you like, it isn't commonplace
Even if it gives you nowhere else to go.

You're on your own. Nothing upholds you here
(Physics apart) but human love and trust;
No grey Jehovah lurks between the stars.

It's sad that astronauts in full career
Should be self-righteous, covetous, unjust,
And crowd our ship with barracks and bazaars.

Nativity

So many times the painters have enscrolled
The scene! The lowly stable thronged with Kings;
Melchior, Gaspar, Balthazar, each brings
His gift of myrrh or frankincense or gold;

The shepherds, having left their flocks in fold,
Gape in amazement at such wondrous things,
And 'Peace on Earth' a choir of angels sings.
It's all familiar, changeless, known of old.

Acknowledgment is owing, all the same,
To those profane assistants, out of sight,
Who give the picture its protective frame:

Partisan outposts up the desert track
Who curse the Star for showing too much light
Yet by their vigilance hold Herod back.

Leonard Mann

Vision

There was nothing there, nothing but a wall
Or, rather, to be quite accurate,
A fragment of a wall. That was all
And in the wall a window set and shut.

Ah, but that was also a strange thing
This fragment stood solid in dark space
And it seemed sure beyond all questioning
The dark window could never frame a face.

A dun diminished light disclosed
The bricks were mortared in a common bond
And sure it was the flushed bricks interposed
Between nothing this side, nothing beyond.

If that window had been thrown up wide
One looking from this side of it would see
Nothing and one looking from the other side
Would see void in extremity.

And sure it was that fragment of a wall
Was not suspended, it was not raised there,
Below, above there was no thing at all,
And on each side nothing, no house, no air.

It was not either in space or time;
It was perhaps a prophecy cast
Up by chaos; it was only a gleam
Or a reflection from an unborn past.

And who was I? Nothing but a thought
About a wall, naked within the void,
A thought within a void and except that,
Nothing, and with that to be destroyed.

And then the window opened suddenly
And in the frame appeared an empty face
By coincidence of eternity
Of time and of eternal space.

To God

Your law is not enough,
You have need to love.

Or the judge of all
Is mummified in law
And with archaic saw
Insults the criminal.

You must be not above
But within all love.

Or fate lays waste the sky,
The sun fails of life's fire,
The moon loses desire,
The stars shine but to die.

Beauty is the stuff
And the soul of love.

Or loveliness is nowhere,
By bloom is the rose defiled,
Innocence kills the child,
The bird falls down the air.

Life is in love
And may not rove.

Or the body is a hearse
In which the soul makes no stir,
The world is a sepulchre,
A graveyard the universe.

In me there is love,
Love you have need of.

Billy Marshall-Stoneking

Sky

The guardians of the circumcision ceremony
live in the constellation of Scorpio, and
turn the sky over every night.
The sky is a shell.
The Milky Way, a creek
of gleaming stones.
The Southern Cross is
the footprint of the wedge-tailed eagle,
and mushrooms are fallen stars.
The sun is a woman,
moving by different paths
between winter and summer.
And Jupiter, the dog,
hunts with Saturn, who brings
bush tucker back to Venus.
There are two moon men—
an old man and his son—who once
lived in the mountains;
the father is so large,
if you saw him, there would be
no room for anything but fear.
The son persuades his father
to stay in camp. Some nights he stays
with him there.
If the father were allowed to rise
his light would blind the world.
And once, after the whitefellas came,
the people from wilarata side
saw Jesus in the clouds.

Philip Martin

An English Martyr

Tower of London, 1535

Now winter and all death are past.
Returning birds, returning leaves
And sun returning: can it be
That any heart this morning grieves?

And surely none grieves less than mine
Which in an hour shall beat no more.
O men unborn who, looking back,
Count me dishonoured and most poor,

I need no tears: I leave behind
All change of seasons and of men,
The pomp and treachery of kings,
Corruption, agony and sin.

The fine and running sand that marks
The hour sifts down within the glass.
A little longer and it's spent:
So lightly may this spirit pass.

And warmth of sunlight falling still
Upon this hewn, this bitter stone,
I take as greeting from my Lord
And rise up joyful to be gone.

A Sacred Way

The God his father preached was a harsh God.
Then as he grew to manhood he must face
Desire as horror: the female legs gaped wide,
The whirlpool sucked him to its hell-black centre.
But there was no escape from womankind,
And past the middle of life the dark wood thinned,
Sunlight among the limbs, and overhead
Leaves swaying in wind the hair of dancers.
He found the woman in himself, and found
In every woman he embraced the earth:
It was from her he came and he would soon
Re-enter her. His father's God was dead
Long since. It was a goddess whom he served.
She spread dark honey on his lips. They sang.

Mudrooroo

From *The Song Circle of Jacky*

Song 22

The road is barred by temples and churches,
My Lord, Jacky hears your call and tries to run:
Prophets and gurus trip his feet and shout:
Your voice is lost in the din of ceaseless babble.
His soles are sore, his knees bare and bloody,
He limps on—but his desire for unity
Dies in a host of faiths screaming TRUTH.
At last, the door of love—bearing too many locks:
Krishna and Allah, Jesus and Kali, Mother and Father:
Poor Jacky sitting and crying in regret and frustration.

Les Murray

An Absolutely Ordinary Rainbow

The word goes round Repins, the murmur goes round Lorenzinis,
at Tattersalls, men look up from sheets of numbers,
the Stock Exchange scribblers forget the chalk in their hands
and men with bread in their pockets leave the Greek Club:
There's a fellow crying in Martin Place. They can't stop him.

The traffic in George Street is banked up for half a mile
and drained of motion. The crowds are edgy with talk
and more crowds come hurrying. Many run in the back streets
which minutes ago were busy main streets, pointing:
There's a fellow weeping down there. No one can stop him.

The man we surround, the man no one approaches
simply weeps, and does not cover it, weeps
not like a child, not like the wind, like a man
and does not declaim it, nor beat his breast, nor even
sob very loudly—yet the dignity of his weeping

holds us back from his space, the hollow he makes about him
in the midday light, in his pentagram of sorrow,
and uniforms back in the crowd who tried to seize him
stare out at him, and feel, with amazement, their minds
longing for tears as children for a rainbow.

Some will say, in the years to come, a halo
or force stood around him. There is no such thing.
Some will say they were shocked and would have stopped him
but they will not have been there. The fiercest manhood,
the toughest reserve, the slickest wit amongst us

trembles with silence, and burns with unexpected
judgements of peace. Some in the concourse scream
who thought themselves happy. Only the smallest children
and such as look out of Paradise come near him
and sit at his feet, with dogs and dusty pigeons.

Ridiculous, says a man near me, and stops
his mouth with his hands, as if it uttered vomit—
and I see a woman, shining, stretch her hand
and shake as she receives the gift of weeping;
as many as follow her also receive it

and many weep for sheer acceptance, and more
refuse to weep for fear of all acceptance,
but the weeping man, like the earth, requires nothing,
the man who weeps ignores us, and cries out
of his writhen face and ordinary body

not words, but grief, not messages, but sorrow,
hard as the earth, sheer, present as the sea—
and when he stops, he simply walks between us
mopping his face with the dignity of one
man who has wept, and now has finished weeping.

Evading believers, he hurries off down Pitt Street.

The Barranong Angel Case

You see that bench in front of Meagher's store?
That's where the angel landed.
What? An angel?
Yes. It was just near smoko time on a sale day.
Town was quite full. He called us all together.
And was he obeyed?
Oh yes. He got a hearing.
Made his announcement, blessed us and took off
Again, straight up.
He had most glorious wings . . .
What happened then?
There were some tasks he'd set us
Or rather that sort of followed from his message.
And were they carried out?
At first we meant to,
But after a while, when there had been some talk
Most came to think he'd been a bit, well, haughty,
A bit overdone, with those flourishes of wings
And that plummy accent.
Lot of the women liked that.
But the men who'd knelt, off their own bat, mind you,
They were specially crook on him, as I remember.

Did he come again?
Oh yes. The message was important.
The second time, he hired the church hall,
Spoke most politely, called us all by name.
Any result?
Not much. At first we liked him.

But, after all, he'd singled out the Catholics.
It was their hall. And another thing resented
By different ones, he hadn't charged admission.
We weren't all paupers, and any man or angel
With so little regard for local pride, or money,
Ends up distrusted.

Did he give up then?
Oh no. The third time round
He thought he had our measure. Came by car,
Took a room at Morgan's, didn't say a word
About his message for the first two days
And after that, dropped hints. Quite clever ones.
He made sure, too, that he spoke to all the Baptists.
I'll bet that worked.
You reckon? Not that I saw.
We didn't like him pandering to our ways
For a start. Some called it mockery, straight out.
He was an angel, after all. And then
There was the way he kept on coming back
Hustling the people.
And when all's said and done
He was a stranger. And he talked religion.

Did he keep on trying?
No. Gave us away.
Would it have helped if he'd settled in the district?
Don't think so, mate. If you follow me, he was
Too keen altogether. He'd have harped on that damn
message
All the time—or if he'd stopped, well then
He'd have been despised because he'd given in, like.
He'd just got off on the wrong foot from the start
And you can't fix that up.

But what—Oh Hell!—what if he'd been, say, born here?
Well, that sort of thing's a bit above an angel,
Or a bit below. And he'd grow up too well known.
Who'd pay any heed to a neighbour's boy, I ask you,
Specially if he came out with messages?
Besides, what he told us had to do with love
And people here,
They don't think that's quite—manly.

The Future

There is nothing about it. Much science fiction is set there
but is not about it. Prophecy is not about it.
It sways no yarrow stalks. And crystal is a mirror.
Even the man we nailed on a tree for a lookout
said little about it; he told us evil would come.
We see, by convention, a small living distance into it
but even that's a projection. And all our projections
fail to curve where it curves.
                                        It is the black hole
out of which no radiation escapes to us.
The commonplace and magnificent roads of our lives
go on some way through cityscape and landscape
or steeply sloping, or scree, into that sheer fall
where everything will be that we have ever sent there,
compacted, spinning—except perhaps us, to see it.
It is said we see the start.
                                    But, from here, there's a blindness.
The side-heaped chasm that will swallow all our present
blinds us to the normal sun that may be imagined
shining calmly away on the far side of it, for others
in their ordinary day. A day to which all our portraits,
ideals, revolutions, denim and deshabille
are quaintly heartrending. To see those people is impossible,
to greet them, mawkish. Nonetheless, I begin:
'When I was alive—'
                              and I am turned around
to find myself looking at a cheerful picnic party,
the women decently legless, in muslin and gloves,
the men in beards and weskits, with the long
cheroots and duck trousers of the better sort,
relaxing on a stone verandah. Ceylon, or Sydney.
And as I look, I know they are utterly gone,
each one on his day, with pillow, small bottles, mist,
with all the futures they dreamed or dealt in, going
down to that engulfment everything approaches;
with the man on the tree, they have vanished into the Future.

The Chimes of Neverwhere

How many times did the Church prevent war?
Who knows? Those wars did not occur.
How many numbers don't count before ten?
Treasures of the Devil in Neverwhere.

The neither state of Neverwhere
is hard to place as near or far
since all things that didn't take place are there
and things that have lost the place they took:

Herr Hitler's buildings, King James' cigar,
the happiness of Armenia,
the Abelard children, the Manchus' return
and there with the Pictish Grammar Book.

The girl who returned your dazzled look
and the mornings you might have woke to her
are your waterbed in Neverwhere.
There shine the dukes of Australia

and all the great poems that never were
quite written, and every balked invention.
There too are the Third AIF and its war
in which I and boys my age were killed

more pointlessly with each passing year.
There, too, half the works of sainthood are
the enslavements, tortures, rapes, despair
deflected by them from the actual

to beat on the human-sacrifice drum
that billions need not die to hear
since Christ's love of them struck it dumb
and his agony keeps it in Neverwhere.

How many times did the Church bring peace?
More times than it happened. Leave it back there:
the children we didn't let out of there need it,
for the Devil's at home in Neverwhere.

At Min-Min Camp

In the afternoon, a blue storm walloped and split
like a loose mainsail behind us. Then another
far out on the plain fumed its corrugated walls.

A heavy dough of cloud kept rising, and reached us.
The speeding turbid sky went out of focus, fracturing
continually, and poured. We made camp on a verandah

that had lost its house. I remembered it: pitsawn pine
lined with newspaper. People lived on treacle and rabbit
by firelight, and slept under grain-bag quilts there.

It was a lingering house. Millions had lived there
on their way to the modern world. Now they longed for and feared.
It had been the last house, and the first.

Dark lightnings tore the ground as we ripped up firewood
and when the rain died away to conversation, and parted
on refreshed increasing star-charts, there arose

an unlikely bushfire in the ranges. The moon leaped from it,
slim, trim in perfect roundness. Spiderwebs palely yellow
by firelight changed sides, and were steel thread, diamante.

Orange gold itself, everything the moon gave, everywhere
was nickel silver, or that lake-submerged no-colour
native to dreams. Sparse human lights on earth

were solar-coloured, though: ingots of a homestead,
amoebae that moved and twinned on distant roads
and an unfixed anomaly, like a star with land behind it.

We were drinking tea round a sheet-iron fire on the boards
bearing chill on our shoulders, like the boys who'd slept
on that verandah, and gone to be wandering lights

lifelong on the plains. You can't catch up to them now
though it isn't long ago: when we came round from the Rift Valley
we all lived in a small star on the ground.

From the Rift we also carried the two kinds of fear
humans inherit: the rational kind, facing say weapons
and the soul's kind, the creeps. Awe, which warns of law.

The two were long bound together, in the sacred
cultures of fright, that called shifting faces to the light's edge:
none worse than our own, when we came dreaming of houses.

Then the sacred turned fairytale, as always. And the new thing,
holiness, a true face, constant in all lights,
was still very scattered. It saved some. It is still scattered.

Many long for the sacred lights, and would renew their lore
in honoured bantustans—no faery for the laager of the
  lagerphone—
but they are unfixed now, and recede, and suddenly turn pale as

an escaped wife dying of a dread poem. Or her child
who sniffs his petrol, and reels like a shot kangaroo:
something else, and not the worst, that happens in a shifting light.

Holiness is harder to inhale, for adventure or desperation.
It cleanses awe of fear, though not of detailed love,
the nomads' other linkage, and maps the law afresh with it.

We left that verandah next day, and its ruined garden
of wire and daylilies, its grassy fringe of ancient pee scalds
and travelled further west on a truck that had lost its body.

John Shaw Neilson

Love is a Fire

It was a rippling day in rising spring.
Soft was the grass: I could not hear your feet.
The world had crazy gone a-blossoming,
Nor lacked a flower to make its joy complete.
You walked close to the barley by the wheat.
The clouds were gone, the red sun reached up higher.
Love, love was nigh, and love it was a fire.

After the battle some will sing a psalm.
The storm will cease and every wind be still.
All those who sat in strife shall know a calm
To rest in a deep valley or a hill.
But love it resteth not nor ever will:
It hath no end, and who shall call it sire,
From out whose womb came Love that is a fire?

The Worshipper

What should I know of God?—he lives so far
In that uncanny country called the blue.
Sweetheart, I cannot worship moon or star,
I worship you.

I shall have miracles of light above.
My church will be an acre of green spring,
And while I pray I'll see the world you love
Still blossoming.

I shall be lifted with the scent of air
And the strong sun will wash my doubts away.
You will be near me when I go to prayer
To hear me pray.

Surely God was a Lover

Surely God was a lover when He bade the day begin
Soft as a woman's eyelid—white as a woman's skin.

Surely God was a lover, with a lover's faults and fears,
When He made the sea as bitter as a wilful woman's tears.

Surely God was a lover, with the madness love will bring:
He wrought while His Love was singing, and put her soul in the Spring.

Surely God was a lover, by a woman's wile controlled,
When He made the Summer a woman thirsty and unconsoled.

Surely God was a lover when He made the trees so fair;
In every leaf is a glory caught from a woman's hair.

Surely God was a lover—see, in the flowers He grows,
His love's eyes in the violet—her sweetness in the rose.

He was the Christ

Our laws, the wisest, haste to die:
Our creeds like idle tales are told:
The loving heart, the lips that bless,
The shadowy centuries make not old.

This life, that ever runs to pain,
He felt it all: its rise and glow,
The bitterness, the ache, the toil,
All that the moving myriads know.

He drew no sword; but all men's swords
Grew redder in the blood-red years:
—Only the hope, that would not die,
Shone tremulous in a world of tears.

The white mist dances in our eyes;
But still, in every age and land,
His heart beats for the little child,
He writes of mercy on the sand.

Song for a Sinner

When you go underground with all your airs,
Your kindly lies and your ridiculous prayers,
You shall not ever fear to face again
The strong man's rage, the woman wild with pain
Nor song nor sigh will beat upon your brain.

The world will mourn you neither less nor more
Than all the pawns who played the game before;
The lover-lad will kiss his love anew,
The water-birds will have their dance to do,
And the rude Spring will gallop over you.

The men who make will match the men who mar,
The eye unsatisfied will seek a star;
Your visitor the worm will speak you fair,
The bride will tremble and the child will stare,
And the red Summer will ride everywhere.

Schoolgirls Hastening

Fear it has faded and the night:
  The bells all peal the hour of nine:
The schoolgirls hastening through the light
  Touch the unknowable Divine.

What leavening in my heart would bide!
  Full dreams a thousand deep are there:
All luminants succumb beside
  The unbound melody of hair.

Joy the long timorous takes the flute:
  Valiant with colour songs are born:
Love the impatient absolute
  Lives as a Saviour in the morn.

Get thou behind me Shadow-Death!
  Oh ye Eternities delay!
Morning is with me and the breath
  Of schoolgirls hastening down the way.

To the Untuneful Dark

All beauty falls to you:
Early white love,
The Reason that draws for us
Splendours blazing above,
The lightning of dancers,
All the journeys of Song;
Ah, well I know you—
You speak too long,
Out of you came God.

Into you fall tenderness
Coated in prayer,
And the red heart of those
Seeking the fair,
All those whose gladness
Puts them upon the blue;
Ah, well I know you—
Who so angry as you?
Out of you came God.

The Crane is My Neighbour

The bird is my neighbour, a whimsical fellow and dim;
There is in the lake a nobility falling on him.

The bird is a noble, he turns to the sky for a theme,
And the ripples are thoughts coming out to the edge of a dream.

The bird is both ancient and excellent, sober and wise,
But he never could spend all the love that is sent for his eyes.

He bleats no instruction, he is not an arrogant drummer;
His gown is simplicity—blue as the smoke of the summer.

How patient he is as he puts out his wings for the blue!
His eyes are as old as the twilight, and calm as the dew.

The bird is my neighbour, he leaves not a claim for a sigh,
He moves as the guest of the sunlight—he roams in the sky.

The bird is a noble, he turns to the sky for a theme,
And the ripples are thoughts coming out to the edge of a dream.

The Vassal

We wait as the trees in the forest, how soon to be thinned;
This vassal has pity—I cannot speak ill of the wind.

He slew the young lamb and the mother, he burnt up and ruined
the hay,
But once at [the edge of] the midnight he caught all my sorrows
away.

The sorrows were thick at the fireside and thick on the floor;
They would not fly out at the window nor look at the door.

I rose, I implored the good vassal, so urgent was I;
He took all my sorrows and scattered them out to the sky.

God is so weary in counting up all who have sinned;
His vassal I know well—I cannot speak ill of the wind.

This quiet little vassal who cools in the joy of the tree,
He roars at the midnight, he shakes on the face of the sea.

He put my despair on the waters, he lifted it clean;
He ran to the edges of heaven, where no man has been.

We wait as the trees in the forest, how soon to be thinned;
This vassal has pity—I cannot speak ill of the wind.

Geoff Page

Country Nun

In a cafe under a lazy fan
she talks with her brother,
the breath of cows upon him,
a line of sun and hat across his brow.
Wimpled above the steak and peas,

she drifts away/drifts back,
floating as she did
in cowfields of their childhood,
lingering on the few books in the living room,
always last to the pool.

From rough-sawn walls
beyond the memory of decision
she moves through knee-high pastures
to a convent gate
farewell.

Soon now
he will need to walk her back,
feeling her lift already
towards the pure insistence
of the bell.

Broken Ballad

The faithful sailing
In their graves
Sing praises to the Lord

As doubters restless
Row by row
Grow leaner and more bored.

Believers patient
Underground
Await the lifting kiss

And sceptics stare
Into the stone—
They had expected this.

Equally
The planet swings them
Round a failing sun

Till doubt and faith
Are frozen iron
And metaphors are gone.

My Mother's God

My mother's God
has written the best
of the protestant proverbs:

you make the bed
you lie in it;
God helps him

who helps himself.
He tends to shy away from churches,
is more to be found in

phone calls to daughters
or rain clouds over rusty grass.
The Catholics

have got him wrong entirely:
too much waving the arms about,
the incense and caftan, that rainbow light.

He's leaner than that,
lean as a pair of
grocer's scales,

hard as a hammer at cattle sales
the third and final
time of asking.

His face is most clear
in a scrubbed wooden table
or deep in the shine of a

laminex bench.
He's also observed at weddings and funerals
by strict invitation, not knowing quite

which side to sit on.
His second book, my mother says,
is often now too well received;

the first is where the centre is,
tooth for claw and eye for tooth
whoever tried the other cheek?

Well, Christ maybe,
but that's another story.
God, like her, by dint of coursework

has a further degree in predestination.
Immortal, omniscient, no doubt of that,
he nevertheless keeps regular hours

and wipes his feet clean on the mat,
is not to be seen at three in the morning.
His portrait done in a vigorous charcoal

is fixed on the inner
curve of her forehead.
Omnipotent there

in broad black strokes
he does not move.
It is not easy, she'd confess,

to be my mother's God.

Curtains, Death and Me

Staying at my
Catholic friend's
I sleep beside a wall of God,
speaking in the pure scholastic
versions of the Word.

Foregathered on those
bookshelves there,
pressed between the acid pages,
the truths by now are hardening
and yellow at the edges.

The cities built
between hard covers,
those towers with their transparent floors,
are raised to heaven from one premiss
and nailed with metaphor,

the physics part
of metaphysics.
They sport their sharp antitheses.
The spines are looking out in rows
at curtains, death and me.

Each day I browse
the shelves at random,
observing where the pages fall.
The nights are counting off and God
survives me on the wall.

Peter Porter

Who Gets the Pope's Nose?

It is so tiring having to look after the works of God.
        The sea will run away
        From martyrs' feet, gay
Dissipated Florentines kiss tumours out of a man's head,
Scheduled liquefactions renew saints' blood,

In Andean villages starved Inca girls
        Develop the stigmata,
        Dying dogs pronounce the Pater
Noster on the vivisection table, the World
Press report trachoma'd eyes that drip wide pearls.

All investigated, all authenticated, all
        Miracles beyond doubt.
        Yet messengers go in and out,
The Vatican fills up with paper. The faithful
Work for a Merchant God who deals in souls.

Was there ever a man in Nazareth who was King of Kings?
        There is a fat man in Rome
        To guide his people home.
Bring back the rack and set the bones straining,
For faith needs pain to help with its explaining.

Fill a glass with water and gaze into it.
        There is the perfect rule
        Which no God can repeal.
Having to cope with death, the extraordinary visit,
Ordinary man swills in a holy sweat.

And high above Rome in a room with wireless
        The Pope also waits to die.
        God is the heat in July
And the iron band of pus tightening in the chest.
Of all God's miracles, death is the greatest.

The Old Enemy

God is a Super-Director
who's terribly good at crowd scenes,
but He has only one tense, the present.

Think of pictures—
Florentine or Flemish, with Christ
or a saint—the softnesses of Luke,
skulls of Golgotha, craftsmen's
instruments of torture—everything is go!
Angels are lent for the moment,
villains and devils are buying Hell
on HP, pain is making faces.
In the calmer sort of painting,
serenely kneeling, since they paid for it,
the donor and his family keep the clocking now.
They say, Lord, we know
Lazarus is king in Heaven
but here in Prato it would be death to trade:
the death of God requires a merchant's dignity
and so they tip their fingers in an arch
that runs from Christ's erection
to a *Landsknecht* leaning on his arquebus.
Those centuries were twice the men
that MGM are—God loves music
and architecture, pain and palm trees,
anything to get away from time.

Looking at a Melozzo da Forlì

And in this instance we think of you, God,
You beard above all things,
Canceller of every fact except death,
Looking down on your grand intercession,
Orthodox, like the artist's vision,
Helpless helper of time and promise.

But we do not get closer to love.
The angel's admonitory finger
And the lily of greeting tell Mary only
That the clock in her womb is ticking,
That she will come sooner to sorrow.

And I can see too in the structures
Of church and family another death.
We are entered by the spirit
And thereafter comes such rich despair—
Sermons of the penis, oddities by the seashore
Where towns have sunk, letters lost
In the mumblings of a drunken alphabet.

What is Mary kneeling on? A yoke,
A box for Miss Plath's mad bees,
A stiff pew for a Protestant Sunday?
In one revolution her body shows
Disquiet, reflection, inquiry, submission, merit.

These shapes Melozzo put on a wall
Fade like the dove-voiced poet
Into a high wood of darkness.
From his flat-bottomed cloud, God observes
Earthly love and sadness, saying
After all, this is only a language of gestures.

Yes Mary, you are an actor in a play
Whose dénouement is now to be spoken.
I rehearse the lines myself to your angel—
The action is beginning, blessèd is the Virgin
Who shall be the mother of death.

An Angel in Blythburgh Church

Shot down from its enskied formation,
This stern-faced plummet rests against the wall;
Cromwell's soldiers peppered it and now the death-
    watch beetle has it in thrall.

If you make fortunes from wool, along
The weeping winter foreshores of the tide,
You build big churches with clerestories
    And place angels high inside.

Their painted faces guard and guide. Now or
Tomorrow or whenever is the promise—
The resurrection comes: fix your eyes halfway
    Between Heaven and Diss.

The face is crudely carved, simplified by wind;
It looks straight at God and waits for orders,
Buffeted by the organ militant, and blasted
    By choristers and recorders.

Faith would have our eyes as wooden and as certain.
It might be worth it, to start the New Year's hymn
Allowing for death as a mere calculation,
    A depreciation, entered in.

Or so I fancy looking at the roof beams
Where the dangerous beetle sails. What is it
Turns an atheist's mind to prayer in almost
Any church on a country visit?

Greed for love or certainty or forgiveness?
High security rising with the sea birds?
A theology of self looking for precedents?
A chance to speak old words?

Rather, I think of a woman lying on her bed
Staring for hours up to the ceiling where
Nothing is projected—death the only angel
To shield her from despair.

The Unlucky Christ

Wherever they put down roots
he will be there, the Master-Haunter
who is our sample and our
would-be deliverer. Argue this—
there were men before him,
as there were dreams before events,
as there is (or perhaps is not)
conservation of energy. So he
is out of time but once stopped here
in time. What I am thinking
may be blasphemy, that I
am like him, one who cannot
let go of unhappiness, who has
come closer to him through suffering
and loathes the idea. The ego now,
that must be like a ministry,
the sense of being chosen among men
to be acquainted with grief!
Why not celebrate instead
the wayside cactus which enriches
the air with a small pink flower,
a lovely gift to formalists?
Some people can take straight off
from everyday selfishness to
the mystical, but the vague shape
of the Professional Sorrower
seems to interpose when I try

such transport. The stone had to roll
and the cerements sit up
because he would have poisoned
the world. It has been almost possible
to get through this poem without writing
the word death. The smallest
of our horrors. When they saw him
again upon the road, at least they knew
that the task of misery would be
explained, the evangelical duty
properly underlined. Tell them
about bad luck, he said,
how people who get close to you
want to walk out on you,
tell them they may meet one person
even more shrouded than themselves.
Jesus's message at Pentecost
sounded as our news always does,
that there is eloquence and decency,
but as for happiness,
it is involuntary like hell.

Craig Powell

The Water Carrier

Yet willow shadows brush
the hill   the light
goes on existing round him   while

his being there brings into wonder what
is given   what is exacted   He
is imagined   say   or moulded to

a solitude   a Chinese
water carrier who has followed me
for years at least   stooped and offering

what must be thought of as
the real   the ungainly
water gristling the tongue

to speak of water   How is
a mountain borne
in a brushline over him

lighter than rivers floating to
the sun?   If this has not been lived
what can there be for saying?   yet the saying

is all that will become   the willow
shade   the water
and upon the mountain

presences   those ancestors who know
how much forgiveness
we require   although I have never

seen them look this way
Eternal absence   what shall I inhabit?
what unity?   I am writing

no   not of songs even   but the Chinese
water carrier who is following me . . .
all the time moving in some other direction

Jennifer Rankin

After Meditation

I turned my neck to the side
and this was a thousand years.

The creek floods.
For seconds the world is surface and depth
while, at the centre, a yabbie
ripples the creek for a thousand years.

Elizabeth Riddell

The Memory

The memory is of grass like a green pond
And of the scent of melons between a drought and the rain
It was autumn and the tides were always going out
And all the moons were yellow.

It was a gentle time without rage or anxiety
As we waited for the flames to die
And for the wax to crust on the altar
And for the last petals to lie
On the marble and gilt
And for the singing to end
And for the prayers to fail, again.

Ecce Homo

Ecce Homo. Here he stands
With face uplifted to the skies
And waits for the light venomous rain
To scorch his eyes.

Behold the man, the simpleton
As innocent as grass and trees.
He does not understand his place
Is on his knees.

Ecce Homo. See him raise
His rusty sword against a ride
Of fissioned witches. He has still
His idiot pride.

Nigel Roberts

Reward / for a missing deity

maybe / yr on sabbatical
maybe yr in the dunny / reading yr reviews
maybe yr in the Pacific / on a Women's Weekly Cruise
& maybe
yr preparing a statement for the six o'clock news
that perhaps
you were trapped in a ski hut / by an avalanche
of Betty Hutton

maybe / yr a war criminal / farming in Chile
maybe yr tapping phones / of subscribers to Dial a Prayer
maybe yr breaking that record / buried alive
12 ft underground
maybe yr weeping / in Farmer's Lost & Found
or perhaps
yr taking an angel out to lunch

maybe / yr in conference
or hitting off from the Club House tee
maybe yr demanding / a fat personal appearance fee
maybe yr on strike / & wont accept / arbitration
maybe yr being impeached / for yr crook administration
or perhaps
yr competing / in a Twist Marathon / on Taiwan

maybe / yr doing / In Service Training
maybe yr delivering newspapers
to pay yr way through Uni
maybe yr doing / Pestilence & Famine / I
& Destruction II
maybe yr on safari / collecting / for a private zoo
or perhaps
you farted / & very quickly / left the room

maybe / yr a casualty / of future shock
maybe yr in the mountains / plotting revolution
& a second coming
maybe yr the phantom of the opera / alone in the box
maybe yr wanking over Japanese woodcuts
of geishas sucking cocks
or perhaps
yr being interviewed by Frost / Fantastic / or Hef.

maybe / yr establishing an alibi
maybe yr being / held / incommunicado
maybe yr holed up / in Chicago
with a contract on yr head
maybe yr loneing it in Denver
in boxcars / boxcars boxcars
maybe you've been transferred to another branch
maybe yr in Paekakariki
maybe yr in Nimbin
maybe yr walking / nicotine desperate / up the road
& perhaps
yr going to be back in five minutes

But
& most probably
i would think—
you were horribly scarred
in a laboratory accident / &
yr too sensitive / to
show yr face.

Roland Robinson

Invocation

Last night I spoke to you from where I lay
under a banksia tree at the top of the gully.
Come, I said, out of the distances of cliffs
in the mists of after sunset, out of the gully
filled with the light of the soaring insects, come
out of the saplings standing against the sea,
against the mountains of clouds, come from the south
with cold first breath. And then, above the sea
appeared the pale gold evening star, swimming
with its smaller companion into the trailing rain.
Come, I said, and you stirred the trees and wreathed
the smoke of my fire and came in the night with rain.

The Curlew

My doorway frames the stars, the low
dark ridge from where the curlew cries.
The cold stars pierce me through. This is
my vision as the midnight dies.

I hear the wind, the grass. The cold
stars burn. I have put off this husk:
become that bird, a voice that first
cried to me through the amber dusk.

This is my vision, that this will,
love, and desire, shall find their peace,
slip from the ache of flesh and cry
out under stars in pure release.

This my devotion where I lie
sleepless within the siding-shed,
hearing that bodiless cry on cry:
Man has no where to lay his head.

The Sermon of the Birds

Related by Alexander Vesper

I was clearing thirty or forty acres once
out on the western range near Nightcap Mountain.
And as I was working I heard a gathering of crows
singing out in a jungle gully. Their clamorous cries
drawed the attention of all the other birds,
jackass and butcher-bird, soldier-bird, sparrow-bird,
scrub-robin, magpie, the black and the white cockatoo,
they all flew down to the crows in the jungle gully.

And I followed after their clamour, and in the midst
of all the splendid excitement of all the birds,
I heard one fellow was singing above them all.
It was the lyre-bird, the mimic of all the scrub.
And they held this beautiful sermon for half an hour.
The birds would stop and listen awhile, but still
that beautiful voice, the lyre-bird, would keep on singing
and draw them and join them all to a chorus again.

And as I stood there and listened, the Scriptures was
hitting me all the time. That sermon seemed
like the prophecy when Christ shall come and summon
the birds, the valleys and hills, the mountains and ocean
to sing in praise of the grace and the reckoning day,
and the beauty of earth in the splendour that He created.
And I went back and told my people of what I had seen,
of the sermon of praise I heard on the mountain range.

David Rowbotham

God of the Cup and Planet

God of the great and little,
Lord of the universe
And of the steaming kettle
In the kitchen of my house,
Because these are apparent
I believe. I take
In every morning's torrent
That washes me awake
With a spooned-out tinkle
Communion just as huge
As what the stars sprinkle
In all the night's deluge.

God of the cup and planet,
Lord of the white plate
And the moon with the sun on it,
I drink the wine and eat
The piece of bread I cherish.
Is there another choice?
How can belief perish
If blood and flesh rejoice
To be alive by living
Eternity a while,
To the peal of spoons' thanksgiving
And the sight of skies as full?

God of the globe and table,
Lord of the linen cloth
And the dome above the gable,
On the great and little earth
I do believe. Why fashion
A wakeful emptiness
Or just an empty passion
While the ceilings of stars hiss
And the kettle spouts its steaming—
Or doubt, to what avail? . . .
Messiah of my dreaming,
Ghost of the kitchen grail.

Noel Rowe

From *Magnificat*

I

The angel did not draw attention to himself.
He came in. So quietly I could hear

my blood beating on the shore of absolute
beauty. There was fear, yes, but also

faith among familiar things:
light, just letting go the wooden chair,

the breeze, at the doorway, waiting to come in
where, at the table, I prepared a meal,

my knife cutting through the hard skin
of vegetable, hitting wood, and the noise

outside of children playing with their dog,
throwing him a bone. Then all these sounds

dropped out of hearing. The breeze
drew back, let silence come in first,

and my heart, my heart, was wanting him,
reaching out, and taking hold of smooth-muscled fire.

And it was done. I heard the children laugh
and saw the dog catch the scarred bone.

From *Bangkok*

(For Gabrielle)

III

No photographs allowed. No shoes.
You kneel and keep your feet
facing from the elevated shrine. You watch

the angel-figures, how their gold
becomes a dream, a flaming balm,

how their open hands,
facing up against the air, having nothing left
to lose, are poor and therefore unafraid,

so climb the hands

stairs within a barely-breathing waterfall

to see the Buddha's emerald face.

It is the way: each pair of hands
allows another step, another emptiness,
with neither shoes nor camera,
beckons you
more closely
the cool and greening mind.

Philip Salom

Inquiry of the Spirit Body

What shape does it have? Is it opaque and worn inside?
Is it homunculus, pea-shaped and chuckling in the pineal?
A transparent sleeve my body wears like perspiration
or clammy from excitation or from fear?
Stroke my arm and shoulder, then sweep my front's
cascade of skin—surely you feel nothing
extra? Is it seen by changes, by gradations,
like a lover's body browning slowly through summer,
or paling, suddenly, from standing near fluorescence?
Perhaps it sits, an energetic cloud (esoteric I'm afraid)
a held flare, magnesium dancing to the inner eye,
the third . . . Perhaps it floats, stands, sits or slumps down
on the table beside me, my hand pushing through it
as I write. Or does it keep an easy distance like a dog?

Perhaps it is myself on the leash, silver
and a mongrel but for my sixth-sense search,
this dogged sniffing at all corners.
Does it carry scars for wounds not yet received?
I think of those believers whose imagery can dress them
sentimental bodies of Christ, all flogged and white,
the nail-holes slowly changing back, then gone
into hands and hearts of the saintly, or deluded.
Or Buddha's lotus body, or the Bodhidharma thousand armed:
I ache to think I carry those around!

I think of Monroe's fear of ageing
stopping so abruptly, as if the spirit was
a petticoat slipped off, leaving nothing.
The haunting image. I saw an old man at the shop
with dandruff and a face so lined it might have grown
through a string bag (his spirit in string, held in and down
and just too small?) Time can make the heavy body light,
or the light heavy, this difference quite extraordinary.
I go on looking. But as Rumi said centuries ago (stuff him!):
what you are looking for is what is looking.

John A. Scott

Four Sonnets: Theatre of the Dead Starling

1

And the sights of life never dreamed, dear
Lady and gentlemen, for You now I step aside
And music! Tambourines played by the bare
Breasted girls! Life, life, O living to be lost
In arms and charms' Clap-clap 'Huzzah!
Did You hear: Tambourines! Taut flesh to shake
A music' 'Drums with soft hair' 'Awaken then
The Middle Ages' 'Flock of light' 'Ask Harry'
'Tassles!' Jaundiced in the flicker. 'Wake and
Heave-Ho, Fleets In, catch the silent issue'
In arms lost to wings, Man to Starling, else
Where the room gained life. Ceiling thrashed by
Flight and beating (That the walls could promise
Air) had collapsed their life by closing time.

2

  The Starlings, so I believe,
Imitate death, here in the theatre'
And sweeping them again from the aisles,
Their legs of branch, hardened, yes, but winged,
Lord, had You not made them to fly above the earth,
In the open firmament of Heaven?
  That night, clientele, random from solitude,
Picked their way past knees and coats 'and
Gentlemen, for You now I step aside' *Taking*
The Tambourine, my body knowingly upstaging,
I lay in whistles, breasts numb with eyes. At a drumroll
Setting fire to Starlings piled on the boards,
As if applause was forever in the sound of flame,
And Lord, had they stolen death most perfectly?

3

A great blasphemy.
'Cover the pair, my boo-hoo innocent' *Lord*
In sin. Raising himself. 'May wings of
Christ enfold thee'   She fled the stage, as
Through smoke there were wings, not flames,
To obliterate the air. As the theatre doors
Closed. *The pyre consumed. Sweeping the*
Dead Starlings from aisles, brushing them from seats
To the floor, every winged fowl after his kind,
And Lord, did You not see that somehow all was good?
'Sights of life never dreamed' *Protect me from*
Exclamation. Let me, this time, take upon myself the Starling to imitate Christ.
As we 'Bang the drum with a Boom-Boom' 'Melons,
Cheap' 'Sway forward lads' 'Jock, you raper, go!'

4

And the evening,
And the morning were the fifth day. Time
In the perfect circle of imitation moving to rest,
Or to light and the void, which is rest.

Starlings build in the theatre a roof
Of voice beyond dawn. For the season of drama they
Descend and are multiplied by the people.
And Lord, Dear Lady and gentle, teach us to pray.

Out of unrest, trembling, men were upon her,
Swollen with blood. Violence of the word, choose
To see two Starlings mate and be consumed in heat,
For this is the purpose and movement of theatre
To a new beginning. The crowd emerged into night
Streets, and there was amongst them dominion.

Limbo

No more than earth, the poem
is a cage for the exhibition of Christ.
The shadows, so singular in their purpose
—an absolute loyalty; a fawning sentiment—
spill across campus, as if the world
and all its lawns here were expendable
to the flood of night.
And on these lawns,
Manet-groupings pause from an evening class;
still nodding out the vague insect-whirr
of their rhetoric; then saunter to the halls.
In solitude take the priest of their notebooks:
write a holier confession, if owing less
to art.
There is always the problem of how
to teach Hopkins (how with their stubborn
biology of knowledge?) to those so at home
on earth.
Here it might be simpler:
showing light withdraw to the highest place;
how it breaks loose like exquisite birds,
preparing to die for God.

Waltz

(About Hopkins)

For if we concern ourselves with the behaviour
of English birds, turning from the construct
towards morning, we will learn many useless things.
Schools of criticism will emerge to reconcile God
with his unhappy creations. If we insist upon
the exclusiveness of meaning, failing to see how
'bringing together' and 'collapse' might be this
same point, the earth will always return to shadow.

For if 'here' is seen as the point of reflection
and not as the place of observation, might not
the natural world easily tend to destroy itself?
As the 'unlikely' behaviour of this windhover
on a late afternoon, moving with purpose through
the clear pictorial space of the sonnet.

Thomas Shapcott

Singing in Prison

But the man is, as it were, clapped
into jail by his consciousness.
Emerson: *Self-Reliance*

It is not for you, and it is not for myself
that I guide my fingers into the wet clay
and draw it out, shape and turn it
with the easiness of practice and the unease
of command into something neither
for nor against but beyond everything it ever was
or I may ever be, or you, also. I do this
to become a pattern, realizing that to displace
and reanimate is our earliest heritage.

My voice thrusts out with eagerness
into the air's clay and water, my singing
moulds the dull materials: how tangible the air is,
how easy to draw-up the light. In the blue
and gold of each minute, my inadequate materials
centre me, we are all spinning. Creation
is the one word, and I shall articulate it
certainly.
Through the window
the bars crisscross my body with light and cold
where the fingers of day plunge in;
I kneel, I neighbour with the dust motes,
to be pierced, displaced by the steady fingers.
The rainbow drenches me
in its one terrible pattern.

Portrait of Saul

i Jonathan and David

Yes, but to remember them for their love
is to remember them for their youth: laughter,
not a covert whispering; the noisy clatter
of playingfield and bodies so alike they move
in a teamwork: do not suppose that what they give

each other is theirs to hold or withhold. Bitter
and old I watch how they embrace each other
free with the one gift I no longer have.

The strings of David's harp are bars of a cage,
a sour taste corrodes through his sweet song.
I am afraid. The desires of a King
are comfortless: my Palace holds me hostage.
And, if I had him, what then could I, Saul,
do but mortify, condemn, despoil?

ii King Saul

Like the sun in his seasons the shadows around me change
and slide. The constant light stoops to play games.
To be the light but not the sun. It seems
that I am King of dustmotes and a rainbow prism.
I sift fragments. Grass is what I govern.
And there is the greater horror: no change
but constancy. The glint of an iron range
laughs at the insubstance of my realm.

So, when my people cry Hail This! Hail That!
I do not need a mirror to be told
that God has withdrawn. I rule my world
without joy, and do not flinch in my right.
I am sharp—I am sharp as sun on steel.
O light can burn through grass: the fire is real.

iii The Old King Saul

To wake in a chill bed—a face of cold
on the pillow, and my own body-heat
inadequate. To wake alone in fright
denied all comfort is to wake still in the world.
This morning I lay and my heavy body rolled
and squatted upon some panic in my heart:
the panic of a boy so young that I forget
he hides within and will not yet be dead.

No. Panic of an old man knowing he is dead.
Knowing there is no comfort, no warm bed,
no parents wide and sleepy and unafraid
to murmur *come here and be comforted.* There is no word.
Somewhere my son lies where and with whom he will.
I offer nothing. But the past rots in me still.

Jemal Sharah

Reliquaries

Gleaned rubble of the Berlin Wall
now gleams in silver settings: tesserae
of brick and plaster for a socialite.
I thought it was eternal: now I see
in my mind's eye, the way these chips would fall
like snowflakes through cold air;
dust shaken from the Iron Curtain's folds
and fashioned into costly ornaments,
emblems of dissolution and downfall,
like the gold crosses that some people wear.

So potent with suggestion: yet not soaked
with blood, or pocked with shot, or clawed by hands.
These could be flakes from any garden wall—
Europe is peppered with ersatz remains:
think of the journeyman or ageing wife
of Compostela, Rome, Jerusalem,
watching a son, a wife or daughter die—
who, dying, took the bread of their old age.
The dead do not have need of hands, or toes:
how tempting, then, it would have been, to shift
their hungry stomach's pain onto their heart's;
to keep their minds quite blank until they'd cut
away the body's profitable parts
and packaged them for hawking by the road.

These crumbs of brick embodied once the faith
that people could be saved from such a kind
of martyrdom; that vision now becomes
an ornament on an opponent's breast.
The world reforms, another dream is pressed
into a bauble, and a vast regime
circles a woman's neck; still, many a queen
of antiquity is known now for her jewels.
This stone, as much as any hope, deserves
memorial; just as the labourers
and dead wives of a thousand years ago
earned through their poverty their borrowed right
to make their testament to history.

Miserere

After the Trinkhalle, its murals of plague and redemption,
the casino's blind weeping windows of winter,
we were hustled by rain to the cathedral.

There was only time for a glimpse up, through an eyeful
of water, at iron walls against the dusk.
The salvo of rain was choked as the great door shut.

Inside, it was as though breath had been taken
into a lung. Too dark to see the roof.
Or the long stained-glass windows: the figures sullen

and random, a pattern of twigs in ice-black water.
Stone piers supported shadows. Only one
chapel was lit beyond the sombre altar.

Tapers were burning the edge of all that air.
A tray of sand was set to catch
every white blood-drop, every burning prayer.

Drawn to their feeble warmth, we came
upon a ledger, in which were inscribed
those pleas too urgent for the flames:

'My brother's cancer . . .' 'Send her home again . . .'
'Help me endure . . .' 'Let this child remain . . .'
—a litany of pain.

Our slow translation left nothing to say.
The great stone walls in steady exhalation
filled the church with their cold. We made our way

back to the more diffuse dark of the world.

Revelation

It's night, and blackness stretches from the road
behind me, out, right out beyond the world.
Alone, anarchic—I should be at home
but have let the bus bring me to its last stop
here, at the wintry beach. You can just see,
in the dark, the white, collapsing foam
and hear, amid its clamour, depths of peace.

R. A. Simpson

Lachryma Christi

Tears upon my lips?
Here is delight at last.
Think only of the taste;
Weeping has long since past—

Or so I tell myself.
How can a God permit
Volcanoes and ignore
The chaos from each pit?

Perhaps lava and ash
Were meant to shape the ground
Until grape-vines could grow
Beneath the smoke and sound

Of new volcanic needs
That even shadow wine.
The tears of God reveal
An absence of design.

Tunnels

I ask myself to laugh;
My laughter's like a croak.
That needle-point of light
Could be a joke.

My feet in muddy water,
My fingers reading walls,
Like fingers knowing braille,
As someone calls

In me? Out there?
The needle-point of light
Expands—the only kind
Of deity in sight—

It lords my view.
Escape.
And so I dare,
And engine like an ape

Toward another opening
Which blinds with so much fire
After so much dark. Tunnels
Follow tunnels of desire.

The Iconoclast

A Citizen of Constantinople—726 A.D.

I'm told—here's Christ. I can't accept
That circle where my gaze begins
A journey down the regimented stones.
I almost hate Christ against their golden void.

The pagan gods are dead but linger,
Looking from His tessellated face.
He should transcend mosaic, be in my head—
And so I put Him there to reign those walls.

Yet even as I live the face, the eyes,
The holy facets give themselves to stone again.
I'm trapped by walls, a church, the ultimates
Of ice and fire. My Lord in me is dead.

Alex Skovron

From *A Life*

The Moth

There is a knowledge indistinctly heard
Behind all that I know and all I am.
Behind the turning socket of the world

It coruscates like some shade harmony, stirred
In the pedalpoint of sleep: I understand
There is a knowledge. Indistinctly heard

And brief, it splashes into colour, flag unfurled
To sputter in the wind—a wind whose chant
Behind the turning socket of the world

Can barely reach me. Yet the droning whirl
Of mind and circumstance, and entropy, demand
There is a knowledge, indistinctly heard

But true. And I'm committed, I am spider-held
To circumnavigate its soft command
Behind the turning socket of the world.

And as I write these lines, my syllables meld,
The music haunts the shadow of my hand.
There *is* a knowledge, indistinctly heard,
Behind the turning socket of the world.

From *Sleeve Notes*

Strings

Another dawn is slowly inflating,
I sense the shudder (the world is waiting)
Of sky and steeple, of mud and madness,
Continents stretching—I am Atlantis,
I am a worm in the dust of an ember,
I am a king, my life is wine,
I am a snowflake lifting December,
I am a pauper, I am divine.

But what if the mouth and mind of a creature
Can *not* be read by a God above—
Illiterate God no prayer or preacher
Can move: just music and love . . .
And is the sleep of God ever crammed
With nightmare dreams of the souls of the damned—
A chasm of voices waiting immensely?
(*If they should ever rise up against me* . . .)

Before we know it a lifetime scatters,
The highest wisdom straddles a bluff:
Everything matters and nothing matters.
Not to look back is not enough.

Peter Skrzynecki

Pietà

Holding the bruised and yellow face
She watched borders of shadow
Outline the same calmness of eyes once seen
At their table, time and time before.
The covering of thorns had ripped
Against the inscription they'd nailed;
That, too, somehow withstood the wind.

She turned the body sideways,
Putting neck and face into the folds
Of a mantle he'd watched her make.

The words of an old man burned
In her head like a whisper
That once awoke her from sleep:

And desert winds of Egypt were seven horsemen
With flaming swords.

Vivian Smith

The Other Meaning

There is another meaning here—in birds
and trees, in love and grief,
in the fall of the blown leaf
and pain and joy shuffled and dealt like cards
—where thoughts in my stubborn land of pain
travel like water over stone.

But I have failed again: failed love
and failed those simple trees that hold
their brief and formal birds, the standing shade;
my steeple world bereaves its bells.

Even now with winter's first red bird,
the hard incisive light and snow beginning
I have failed: finding the world alive
with pain, and without its other meaning.

Return of the Prodigal Son

Years of terror, in the mud of years,
absent from the self; the self alone;
wounded like an animal and nothing real
but the closed reality of pain

as hard and shut as stone, the thorn, the land—
not even knowing that he took a way;
but only that a change can bring relief—
not even seeing, even knowing what to say

to those who passed him on the furtive road,
or any thought to see his father stand
beneath a palm tree in a fly-blown shade;
unaware of body, face, or hand:

he stumbled on the roadway in the sun,
a mirage or a vision, falling, fell
and broke into the country of his heart
and lay there drinking by its dying well.

O life that beat his head against the road,
O seed, oasis, O consenting heart.
Black with light a fountain wept.
And knowing nothing, knew he must depart.

The Traveller Returns

We do not know if gods preside
but I believe in angels seeing clouds
pierced by rays through pencilled distant slopes.
After slow cathedrals, pilgrim towns
Sydney's violent sky can offer this

moment that catches us still unprepared.

Murillo's dark madonna knew such cloud.

Watching the Pacific lick its samples of gold leaf
I voice once more my disbelief aloud.

From Korea

A cuckoo in Korea called me out
towards the forest near the new hotel,
the light of dawn still tender on the trees.
I heard the bird quite close I couldn't see.

The garden looked alive, alone, itself—
pines propped on crutches, lichen healing stones,
and water in a pool that plopped and flopped.

The Silla hills of Kyongui at dawn,
the clipped grass of ancient tumuli—
we need such conversations with the dead,
or if not conversations presences,
the sense of clear proportions cut in stone.

Sokkuram's Buddha calls its pilgrims up
the long path that leads into its cave
and has done so for fourteen hundred years.
The apostles could be Gothic effigies.
We look at them through glass among the crowds
who come as tourists not as the devout.

Some are born to faith and fixed belief,
some are born to wander and observe,
and some revere what never can be known.

These hills are older than the tombs they hold,
more ancient than the temples, trees and grass.
The spark of life cannot be held in stone.

The cuckoo keeps its call up in the trees
a moment longer as the day begins
with harder light and lorries on the road.

Vyvien Starbuck

Holy Thursday

Night blows me like a broken limb
into the church.

The walls have been painted white
and new red carpet interrupts
the hedgerows of pews.

I don't like change.

I am told that God is gone—
there is truth in this; the tabernacle
has been stripped back to a shell,
the avid flare of a Roman soldier's helmet.

Candles are trunks of wax riveting the altar.
Something begins.
Words disperse in incense.
I simply do not understand what is going on,
I turn and walk out.

The streetlight is an eye
stilting on a spine of cement;
it haloes the leaves with
the pallor of silver.

Crucifixion

Killing this man
is a work of art.

His cross is cast
in a hillside rut.

The man hangs on the nails—
the nails that—
centring his palm
redirect the flow of blood,
the shank of muscle.

His ribs part like waves—
they cage a little life
measurable in the blood
that spills shadows on the skin.

His body is metal
tarnishing with sweat.
His legs crook in
a flimsy quadrilateral.

Soldiers crouch in the dust like
lepers waiting for a miracle cure,
prophesying death.

The sky drops its shroud folds on the hills.
One hill is hollow, with a stone door.
Full of slabby darkness, it awaits the light.

Peter Steele

A.D. 33

In the end at least the dying was not his
  operation because there were others to take
measures and kicks so all that he need take
  was his time, what there was of it, and get
things in perspective, allowing for the odd
  effect brought on by the angle. At last
he had a posture he need not hold to guide
  idler, fanatic, and the simple poor
devil forked on the avenues of life:
  something in league with his bones, a stand
on nothing but nothing, a blueprint for himself.

  If he left them alone they would leave him alone,
and did for the most part, strolling back to town,
  though a few creaked on in leather and iron
who were foreign and troubled professionally and so
  not creditors of forgiveness when
he asked it in their name. It was only fair
  that he need not forgive himself in the end,
being naked and harmless and lost in the sick ache
  he brought his partisans and healed
with a call something between a drug and a bribe,
  but hung in his peculiar guilt,
private and burning as the swollen lungs,
  afraid that it was and was not the end.

Matins

Out there in darkest Parkville it's a kind
  of animal country. Morning displays—
I thought it was the gardener—someone trotting
  hale and compulsive, barely attached
to four maleficent greyhounds, sleek and dumb.
  He's Bogart or Camus, a bigboned ghost
easing himself and his charges round the block;
  they move as sweetly and as bloody-minded
as if their talent were for treachery,
  not coursing and a would-be kill.

We've traded words on form in wetter days,
   sodden together into comradeship,
but not this morning. I'm praying in his trail,
   a sort of christian and a sort of man,
watching him get between us the police
   the park the children's hospital
the bolted shelter for old derelicts
   and the zoo, that other eden, where
some cruciform and prestidigious monkeys
   hang in the sunlight, and the sombre bears
rove their concrete to sweat out the duration.

To Thomas More

I used to think of you as a figure of style,
   who poised himself with the others for Holbein:
a wit as sharp as God could want, and as blunt
   as the king could bear: a martyr caught
at the edge of death, unable to break the habit
   of making jokes: amused and English,
performing homage to heaven not from a pulpit,
   but in a theatre of the absurd.

Perhaps. But now, though Henry is gone to his place,
   the wrath of the king is still death: and now
that Utopia lingers, just as you guessed, nowhere
   and everywhere: now that children debate
the acting of Scofield, but cannot imagine the point
   of the prayers in the dark, I return to your image,
hardy thinker, gentle speaker, courtier
   of life, a humanist to the death.

I should ask your prayers—by this time, in the light—
   if I did not suppose that a man like you
had remembered me before. I have not seen
   your England or your Heaven, but
I rifle books, make jokes with beasts, endure
   a world whose folly is past all blame,
whose dearness past all praise. Old friend,
   think of me often, speak of me with God.

Covenant

I

Quiet, and quiet, and quiet. I do not know
Whether it speaks of life, or promises death
But night and day I am looking for silence, hungry
For life or death, the yield of light in darkness,
Instead of this pacing blindfold in a cage,
The world a roaring vortex, the past a heap
Of promises lost before their words were out,
And the future a bait shifted about the cage.

Quiet for the bones, the heart, the leaping tongue,
Quiet where deeds are mute or clamorous,
Quiet until the dream of action brims
With hope no longer random, and the dream
Flows over into action. It has taken
The time of mountains shredding into sand
To gauge my kind of waiting, and to teach
The kind of need I have; but I learned, in the quiet.

Dark one, light one, dark as the heart of a flame,
If you can speak in silence, practise your art.
Be a pillar of fire or a pillar of smoke, or the small
Voice of a mountain pouring in sand through an hourglass:
Be grave if you choose, or turn in joy
Like a tree tipped with light, or the first
Wave of the sea, exulting in its planet.
Bright one, dark one, speak to us out of our silence.

Of God and Despair

Scrabbling away at the numbers, telling us all
That what counts is What Counts, and tilting the swill from a
  bucket
Of items that look like the horn's and smell like the street's:
Mumbling the chime between brainsteam and loinstem, seeding
Assurances that silver's full of golden
Promises (not pro's, and not, sure, misses):
Dolling oneself up: glitzing the upflung throat
To meet the gauging confreres, and to faffle
Sweetheart talk, the eye a bitter olive:
Canting one's conversation, nine or noon,

At charming chatterbabble: lipping the sounds
That make for solace, heart-crust caving in:
Stinking, throbbing, yearning for meaning, desiring
The long, blond, irretrievable strand of you,
The vexed, shambling, quickstarted vein of me.

Harold Stewart

Lingering at the Window of an Inn after Midnight

At the very moment when we are moved to utter the Nembutsu by a firm Faith that our Rebirth in the Pure Land is attained solely by virtue of the unfathomable working of Amida's Original Vow, we are enabled to share in its benefits that embrace all and forsake none.

Shinran Shōnin, as quoted by Yuien-bō
in his *Tannishō*

I

At Shirakawa after midnight: still
Leaning upon my upstairs window-sill
In hot oppressive darkness, I despair
Of finding even the faintest breath of air
That might relieve my sweltering distress;
But in this black humidity's excess
No leaf will stir. I wait for sleep in vain,
Though all night long the voice of water rills
In trickles channelled from the Eastern Hills;
Soliloquising in the trench of stone
That runs down every quiet street and lane,
It lulls me with an eloquent undertone,
But not to sleep. In vegetable patches
And flooded plots of rice at side and back,
The frogs that, following their cantor, quack
In flat cacophony till croaking-matches
Rapidly overlap from ditch and drain,
All stop at once; then they start up again.

So by monotonous rote I, too, repeat
Meaningless invocations, bound to fail
For lack of faith; in pious counterfeit
Telling my round of beads to no avail.
Though I recite the words, no help can come
From Jōdo, for my obdurate heart is dumb.
How could a mere six syllables, so frail,
Feeble, and ineffectual, that bear

On speech vibrating briefly through the air
The Name of Amida, alone prevail
Against the infernal forces that assail
My sceptic mind with dread, dismay, despair?
How could a mythic Buddha, as the vower
Of some imagined paradise, have power
To rescue me with all my faults, enslaved
By passionate cravings, darker, more depraved,
In this degenerate age when few are saved?
And yet the Pure Land saints and sages claim,
Despite all rational dispute and doubt,
That sole reliance on the Vow and Name
Can bring miraculous Rebirth about,
Whereby my undeserving self inherits
A Bodhisattva's six perfected merits.

Though Mount Daimonji's shaven crest, for lack
Of summer's farewell Bon-fire, still is black,
Looming against the humid starless night,
On wooded slopes infrequent points of light
Amid its sash of vapour come and go,
Shooting aloft a blurred and baffled glow.
The dark mysterious house across the street
Rises two-storeyed from behind its fence,
Sequestered by the darker garden's dense
Bushes and trees. Why is it in retreat
And closely shuttered, even in this heat?
Under its rustic gateway's shingled eaves
The lantern scattering light among the leaves
Has long gone out. Now suddenly upstairs
A panel sliding open in the night
Shows through a papered lattice-work of squares
The dimly amber glow of candlelight.

On shrubberies below, obscured from sight,
Gardenias have unfurled their ivory-white
Petals in whorls, and so profusely bloom
That climbing up the wall beneath my room,
Eddies of warm delicious perfume creep
In at my window, soon inducing sleep . . .

II

Hours later: in the huge and sultry gloom
A temple bell has tolled with solemn boom:
Its lingering overtones profoundly steep

The distant stillness, where it still resounds.
Again the heavy pole is swung, and pounds
Its tongueless dome, whose bronze vibrations vie
In their sonorous hive, and humming deep
Pervade the hush that holds the earth and sky.
The damp air breathes, lifting the slightest sigh:
A little windbell, hung beneath my eaves,
Instantly rings its lightly trilled reply.

I wake at once out of a lifelong sleep:
My being's inmost solitude receives
A summons that dissolves its sombre spell.
The Heart's reverberations rise and swell
Till lips and tongue spontaneously exclaim:
'Amida Butsu!'—Buddha's sacred Name.
At once I utter my submissive cry:
'Namu Amida Butsu!' in reply.
For while his call commands, I am not here
To doubt or disobey; my thoughts in blind
Confusion can no longer interfere:
Only his Name resounds within the mind,
And he alone is present in his Name.
So, as the Nembutsu is pronounced, I hear
The ineffable Will of Amida behind
The spoken words that momently appear
Out of the soundless Void within, and then
Into its Ground of Silence fade again.
During this call our voices sound the same,
And yet I do not call on him, but he
By my response recalls himself through me.
All his Compassion and Wisdom are enshrined
Within this one Nembutsu. It now bestows
Initiation from the Buddha's Mind
Upon my own, until it overflows
With calm inherent Light. Its sounds endow
My heart with pure and boundless Life that knows
The power of Amida's perfected Vow
To save all beings who but once proclaim
With single-minded Faith his holy Name.
Amida's mercy need not summon twice:
After his first definitive recall,
No one falls back, no doubts remain at all
Of ultimate Rebirth in Paradise;
For which with many callings I express
My heart's devotion, praise, and thankfulness.

III

Although the stifling summer night is far
Too veiled with haze to show a single star,
Just before dawn above the mountain breaks,
A light inside a lonely farm-house wakes.
The dark by gradual shades has been withdrawn
To leave this delicate-tinted transience
Of clouds above Japan, a touch of dawn
Like rose and white hibiscus-flowers that lay
Their heads upon a weathered brushwood fence.
The dull air freshens, barely felt and brief:
I drink its fugitive coolness with relief.
Since Amida's unspoken Name united
With that dry Nembutsu which my will recited,
The same three words, calling far off and faint
Within my heart, answer its vain complaint;
His gracious consolation is transferred,
Embodied audibly, through every word.
My weakness feels the strange resistless strength
Of Faith flow in, that will prevail at length;
While all my restless questions are resigned,
And silence has absorbed the noisy mind.

Already dawn is whitening into day;
But though the early sky is overcast,
Under the clouds the morning light assures
That in this life, or when its night has passed,
However long confusion still obscures,
The radiance beyond will rise at last.
I slide my window open: from this height
Vistas of silent houses stretch away,
Not yet awakened by the clouded light
Over their undulating roofs of grey.
Shōji, removed to catch the cool, disclose
Interiors, which their paper panels screen,
With mats of rush-straw, neatly kept and clean,
Where the natural brown of woodwork shows
Their airy structure: empty, light, serene.
Kitchen gardens, edged with cypress-thickets
Tied to a trellis-fence of bamboo pickets,
Are still deserted: woven cobwebs lie
Like scraps of dusty muslin spread to dry
Over their hedges. No one comes in sight.
A storage loft with mud walls plastered white

And tiles of glistering grey is glimpsed between
Secluding maple-trees that intervene
With fanning branches, delicately green.

Sunk to the eaves among its shrubs and vines,
One isolated dwelling here declines
Into neglect and age, till almost drowned
By waves of verdure, varied in their greens.
A sagging wicket-gate that hangs aslant
Amid the weary straggling fence, whose scant
Staggering bamboo sticks enclose the ground
Still under cultivation, through its gap
Leads to a little field of aubergines,
Their leaves and fruitage dark with purple sap.
There young green sallows by a ditch surround
The fallow patches ready tilled to plant,
And lightly tousled on the returning tide
Of air, reverse their silver underside,
Fitful as faith that vacillates and veers.
But on the foot-hills, where the pines have laid
Upon their slender tilted colonnade
A roof of needles, thatching it with shade,
The tallest pine has stood a thousand years
Above the wooded ridge, with trunk and bough
As strong and straight and steadfast as the Vow
To save all beings. Looking farther down
Each leafy lane and narrow avenue
To where they end in fields beyond the town,
The rounded Kyoto hills, abruptly blue,
Misty with conifers, close in the view.

Randolph Stow

Ishmael

Oasis. Discovered homeland. My eyes drink at your eyes.
Noon by noon, under leaves, my dry lips seek you.

The red earth arches away to gibber and dune.
I shall not return to this uncharted spring.

Antarctic seas work statuary of ice,
and sand-toothed wind, in the hungry waiting country,

raises unseen its pale memorials
to lioness, sphinx and man. These blinding images

I call to mind to mould the mind, inviting
desert and sky to take me, wind to shape me,

strip me likewise of softness, strip me of love,
leaving a calm regard, a remembering care.

Whoever loves you, whoever is loved by you,
speaks from my heart. That said, enough of speaking,

a clean break now. My ghost will not come creeping.
One night for words, and then my tenure ends.

The hawks wheel in the dawnlight, the dawn breeze blows
from the heart of drought, from the hungry waiting country

—and what have I to leave, but this encumbering
tenderness, like gear for ever unclaimed.

From *The Testament of Tourmaline*

Variations on Themes of the Tao Teh Ching

I

The loved land breaks into beauties, and men must love them
with tongues, with words. Their names are sweet in the mouth.

But the lover of Tao is wordless, for Tao is nameless:
Tao is a sound in time for a timeless silence.

Loving the land, I deliver my mind to joy;
but the love of Tao is passionless, unspoken.

Nevertheless, the land and Tao are one.
In the love of the land, I worship the manifest Tao.

To move from love into lovelessness is wisdom.
The land's roots lie in emptiness. There is Tao.

IV

The spaces between the stars
  are filled with Tao.
  Tao wells up
like warm artesian waters.

Multiple, unchanging,
  like forms of water,
  it is cloud and pool,
ocean and lake and river.

Where is the source of it?
Before God is, was Tao.

V

  A smith at work
does not consult the iron.
  Passionless, silent,
he forms it to his pattern.

  Forge-flames leap
in fragile multiplicity,
  changed, renewed
by the breath of the empty bellows.

  What can be empty
yet ever and all replenishing?
  Under the bellows
blazes the world's forge-fire.

VI

Undying, the yielding darkness
no matter can withstand.
The gate of the yielding darkness
is the root of the lasting land.

Out of the gate of the darkness
fountains of being start.
Dark fountains of absence
play in the darkened heart.

VII

The loved land will not pass away.
  World has no life but transformation.
  Nothing made selfless can decay.
The loved land will not pass away.

The grown man will not pass away.
  Body is land in permutation.
  Tireless within the fountains play.
The grown man will not pass away.

VIII

Grown men are water,
  seeking the basins,
  close to the darkness,
  feeding the land.

A mud-brick house
  for a heart of silence,
  yielding in friendship,
  unveiling in speech.

Grown men are water
  and uncontending.
  Rooted in darkness.
  Feeding the land.

X

This is the ideal: to embrace with the whole soul
  the One, and never, never again to quit it.
To husband by will the essence of light and darkness,
  to grow passive and unselfknowing, as if newborn.
Till the doors of perception are cleansed and without distortion,
and knowledge, motive, power become curious noises,
a total wisdom being paid for a total yielding.
  That is the ideal, the ideal of the mystic leader,
the people's fountain, the channel of changeless Tao.

XII

The colours of time blind the eye to timeless colours.
The music of time dulls the ear to timeless music.
The flavours of time spoil the palate for timeless flavours.
The diversions of time dull, blind and spoil the mind.

XVI

Deep. Go deep,
as the long roots of myall
mine the red country
for water, for silence.

Silence is water.
All things are stirring,
all things are flowering,
rooted in silence.

Silence is empire.
Tao is eternal,
flowering, returning,
with water, with silence.

Deep. Go deep
as the blossoming myall.
Silence is lifeblood;
returning, flowering.

XXV

Before the birth of heaven and earth existed
only that absence, changeless, ever-changing,
midwife and womb of all things, having no name,
whose sound in time is nothingness, or Tao.

The guide of man is earth. Earth's guide is heaven.
The guide of heaven is Tao, the empty road,
winding through stars and time to make its end
endlessly at its own last, next beginning.

Great is man. Great is earth. Great is heaven. Great is Tao,
in caves of space, in the all-space of the atom.

XL

There is no going but returning.
Do not resist; for Tao is a flooded river
and your arms are frail.

The red land risen from the ocean
erodes, returns; the river runs earth-red,
staining the open sea.

Before earth was was molten rock, was silence.
Before existence, absence. Absence is Tao.

LXXXI

If my words have had power to move, forget my words.
If anything here has seemed new to you, distrust it.

I shall distrust it, knowing my words have failed.
In the truth of the indwelling Tao there is nothing strange.

Words well and sperm jets, sap mounts and fountains flow
from dark to light, from Tao to the lasting land

that my words commend, whose names are sweet in my mouth.
In the silence between my words, hear the praise of Tao.

Jennifer Strauss

The Anabaptist Cages, Münster

1535

Jan van Leyden, Prisoner:

It is enough that God is with me;
I need no priest.

The Sentence:

And let the bodies of those condemned—
Krettech, Knipperdolling, Jan 'the King'—
Being brought from the place of execution
Be severally hung in iron cages
Wrought to that purpose.
And let the aforesaid cages hang
High on the steeple of St Lambert's,
That being the place of first offending.
And let the people thus remember
What follows of misery and excess
When foolish men puffed up by wicked pride
Despise the just and natural laws
Of God and princes.

The Polygamous Wife:

Brag in the wind, old bones!
Preach in your stinking cage till the trumpets sound
To set to partners in that resurrection dance
Where there'll be neither marriage nor giving in marriage.
Dreaming, we thought you promised with God's voice
Our spirit's freedom, but woke to find
You'd bound us harder than ever before
In marriage and childbed. A prisoner to his cell,
Battering at hateful walls, you entered my flesh.
Sisters in God? Did a brotherly hand
Slash off my friend's head in the market-place
For 'disobedience'? Did you not hear us all
Pray in our hearts with our first martyr
'See to it heavenly Father—if you're Almighty—
That I'm no more forced to mount this marriage-bed.'?
You could say that He answered. I say rather
Let them toll the cages, not the bells,

Let the cages cry to the Sunday city
'Where is God now? Your God? Our God?
Where is God? Is God? Where?

The Priest of St Lambert's:

God in my hands: shall I offer Him then
To a congregation with eyes glazed
By terror and something more—a terrible greed
Unsated by mere symbols of torn flesh?
The Bishop says that God is Love,
The Bishop says God is in the wafer,
The Bishop says the Church is in God:
I would set down God and Church together
For my hands' bones ache with weight
Even as the beams of the church groan
With the spire's burden. Last night
In the chancel I found another crack.
Every night I beg my God
That the great stones fall
And set me free, that the earth
Open, and swallow me whole.
But is it the same unanswering God
He cried to, breaking upon the wheel?
If I spoke my doubts they'd call me
At best possessed and hunt a witch to burn
At worst, corrupt with heresy.
I have seen exorcism, I have been
Shown the instruments of interrogation;
I am too afraid. In dreams the altar rails
Close round to cage me in.
If the Church be the instrument of God
Let Him use it and make an end.

The Girl:

Every night my heart knocks in its cage of ribs.
If it got out, how they'd startle
These grave masters, hitching their pants,
Laying down coins and solemn reflections
On fallen man. Thoughts are like stones.
My lover's hands were gentle, to me at least.
Let them think they have him, rags of flesh,
Snared in their iron cage. I know
I can charm him out. Every night
Between midnight and dawn he sings in my thighs.
They'll not burn me; by day

I creep about in the roots of the city,
By night I have my protectors.
What are beliefs? We might have had children.
In love he'd call me his mouse, his rabbit—
They crunched his bones in the teeth of their traps,
They flayed him living with red-hot tongs.
I vowed the day they set his corpse
To dangle on their 'House of Love'
I'd never think of God again.

1982

The Tourist:

They seem so insignificant there on the steeple,
Quiet as a birdcage after the bird has flown;
Centuries of rain have rinsed the stones of anguish,
If they are crumbling it's not from the workings of blood;
Terrible things are done, now as yesterday.
Leaving through sunlit woods, I watch a hawk
Sweep, hover and strike. Unheard on the wind
The thin wail of whatever small furred thing
Had blundered into the open, natural prey.
Leaving Europe, I pack away a Manichean postcard:
The world as God's cage for heretics.

Andrew Taylor

That Silence

Finally

silence

not even the dumb butting of snow on
dark air
nor the mute
nakedness of twigs waiting for spring
nor the rich earthy music of stone
as in mothering mud
it nuzzles farther from the sun
not even the chatter of stars
nor the slow
ambiguous
basses of clouds mountains ocean
not even
the dazzling song of ice in shrill air
nor a slant
sun's
innuendo

but

behind it all
crouched somewhere in your eyes
darker than love than concentration
that sobbing terrible hymn that you will never hear
curled like an infant
in the black
back of your eyes

somewhere within you
you
oh
so utterly silent

The Invention of Fire

Under every cathedral
there's a spring of pure emptiness
architects and priests search out these springs
wherever they find one a cathedral's built

without cathedrals emptiness would water the land
it would flow through the long wet lashes of grass
and under the massed white and yellow flowers
and under the faint red filmy leaves of spring
and over the sparkling stones and around the roots of trees

it would find out valleys and engrave them
with its own downward crashing capture of light
it would swell into rivers shaded and wept by willows
and join a sea forever empty of boats
forever empty of children playing on its shores
whom it aches to embrace and whose castles only
it could erase

inside each cathedral a fish floats
high in stone air and in a sky of glass
he is the sun's fish dreaming of that spring
and in his eye we swim to his dreamt heaven
around its shores little houses are built
and children clap at the incense of small fires

The Gods

They've put off their wings, their beaked heads
their yellow-fanged muzzles—
they're rocks on a beach in summer
salt frost blistering their skin

They've shrugged up from the sand
startling those children building castles
they're dark despite the sand's dazzle
and they appear to mean to stay
despite the coming and going of the sea
which up till now was also considered a god

They've nothing to say
except for their dark intrusion
they neither caw like crows nor whimper like wolves
they've put off those childish masquerades
for the enduring inertia of rock
being gods

Maybe if the sea were as still as death
we could watch them move
or if the sea would remain as quiet as death
we could hear them breathe

Dimitris Tsaloumas

ΘΕΟΙ

Δύσκολες βέβαια οἱ περιπέτειες τῶν μεγάλων
μὰ δυσκολότερη ἀκόμα ἡ μοίρα τῶν Θεῶν.
Τὰ χρόνια ποὺ βασίλευεν ὁ Οὐρανὸς καὶ οἱ γιοί του
σαπίζανε στὸ Σκοτεινὸ Κατώι, μὲ τὸ δρεπάνι
τόνε μουνούχισε τ' ἀπογέννι του νὰ πάρει τὰ πρωτεῖα.
Μὰ κι αὐτουνοῦ ὁ βίος ἀβίωτος, ὑποψία
καὶ παιδεμός, νὰ καταπίνει τὰ μωρά του
καὶ νὰ τρέχουνε στὰ κατσάβραχα τῆς Κρήτης
νὰ κρύψουν τὸν Ἄνακτα ἀπ' τὸν ἄκαρδο γονιό του.
Ὄχι πὼς πέρασε κι ἐτοῦτος ζωὴ
χαρισάμενη: ἀγώνας κι ἀμάχη, προδοσία
καὶ χαλασμοί, κι ὁ κόσμος νὰ περιμένει χαΐρι.
Ὁ Χριστὸς τὸν πόνεσε τὸν κόσμο, μὰ τὰ ταξίματά του
σκοτεινά, καὶ τὸν καρφώσαν ἀνυπόμονοι οἱ σοφοί.
Μονάχα ὁ τελευταῖος, ὁ τυχερότερος Θεός,
ποὺ πλερώνει τοῖς μετρητοῖς καὶ σοῦ ξοφλάει τὰ χρέη
μὲ πλάκα καὶ κοντύλι, μονάχα αὐτὸς
πέθανε στὸ κρεβάτι του κι ἀναγνωρίστηκε Σωτήρας.
Ἀπὸ τὸν ἄνθρωπο μαθαίνουν ἀκόμα καὶ οἱ Θεοί.

Gods

The adventures of the great are difficult enough,
but far more difficult the fate of the gods.
When Uranus ruled and his sons were rotting
in the Nether World, his last-born took a sickle and
chopped his cock right off, to gain supremacy.
But life proved wretched for him as well: suspicion
and tortured anguish, making him swallow his own offspring
and causing folk to run across the mountain crags of Crete
to hide the Sovereign from his heartless parent.
Not that the Sovereign Zeus has passed his life
in happy ease: conflict, discord and ruinous treachery
with the hapless mortals waiting for justice.
Christ did love this world but promised darkly
causing the wise impatiently to nail him up.
Only the latest, the luckiest god, who pays
in cash and clears your debts with pen and ledger,
only he died in bed, a recognized Saviour.
Even the gods can learn from man.

TRANSLATED BY PHILIP GRUNDY

Urumbula People of Central Australia

The Urumbula Song

The narrowing sea embraces it forever,—
Its welling waves embrace it forever.

The sea, ever narrowing, forever embraces it,—
The great beam of The Milky Way.

Its embracing arms forever tremble about it,—
The great beam of The Milky Way.

Set in the bosom of the sea it stands,
Reverberating loudly without a pause.

Set in the bosom of the sea it stands,
Sea-flecked with drifts of foam.

The tnatantja pole, flecked with drifts of foam,—
The tnatantja pole casts off its foamy covering.

The tnatantja pole strips itself bare like a plain,—
The tnatantja pole untwists and frees itself from its covering.

The tnatantja pole rises into the air,—
The great beam of The Milky Way.

The kauaua pole rises into the air,—
The great beam of The Milky Way.

The great mulga beam rises into the air,—
The great beam of The Milky Way.

It showers sparks like burning mulga grass,—
The great beam of The Milky Way.

The great beam of The Milky Way
Gleams and shines forever.

The great beam of The Milky Way
Casts a flickering glow over the sky forever.

The great beam of The Milky Way
Burns bright crimson forever.

The great beam of The Milky Way
Trembles with deep desire forever.

The great beam of The Milky Way
Quivers with deep passion forever.

The great beam of The Milky Way,
Trembles with unquenchable desire.

The great beam of The Milky Way
Draws all men to itself by their forelocks.

The great beam of The Milky Way
Unceasingly draws all men, wherever they may be.

Translated 1962 from the Aranda by
T.G.H. Strehlow

Vicki Viidikas

Glimpse

We set off to see *Tara Devi*, Star Goddess, in Her niche of a stone temple. A little toy train clacks along rails from Simla, the hills rise and fall emerald and apple green, first stumps of the Himalayas.

Inside the carriage a man from Delhi quizzes us about our origins. His wife sits silent, never introduced, her solemn face nurses the baby, the food bundles, the commitment of marriage . . . Oh the seriousness of her stare, the redness of the *tilak* bleeding over her third eye.

The Delhi man gives us apples and we munch their cool white flesh—everyone stares as if our foreign mouths are different from theirs. We wonder why Indian men choose a world apart from women, why lock them into virtues and set them in the wings?

We get off the train and wind our way up a hill, around forest stumps, stones, bright grass and curly ferns. Ravines zigzag away to cottonwool clouds, a sky like an ocean washing in freedom, forever open . . .

A raw wind tackles our clothes, a red scarf streaks away as clouds break up into ships, beasts, idols. We never make it up that hill but retrace our steps into a small grunky *bazaar*. A black dog wanders past with the matted locks of a holy man in its tail.

Up there the Goddess waits, She glitters in Her cell strewn with fruits and flowers, little lumps of sacred ash. Her benevolence drifts down the hill, She shoots us an enigmatic smile and we catch Her glow inside . . . Glimpse orchards, the gift of living. Oh Goddess give us courage to love this world as whole.

Varanasi (Uttar Pradesh)

The Ganges is the most beautiful river, a gentle waterway with sandflats and shifting lights. Everywhere the *ghats* cannot keep back Her tides: *Ganga Ma*, Goddess of Sweet Water, so abundant that *Shiva* caught Her in His hair to stop Her flooding the world.

Temples poke their conical hats to the sky, the air breathes of people, fish, activity, everything laid out bright like the washing on Her banks; the river floats its life and death simultaneously.

Mauve winter skies streak above water, blue eyes mirror icons and words, vibrant clarities, like corpses on pyres of sandlewood and flame. Varanasi, Benares, Kasi: said to be the oldest civilised city in the world, you suspect that by the way drains ooze along alleys.

With the temples of Heaven and the gateways of Hell, the Ganges is a gentle Goddess, sifting rings of mud with collisions of dust, Her body changing of water like the Dreamtime Serpent. Wandering Her banks with their signatures of ash, I saw the wood dead carve themselves—for an instant, a source of hope.

Saint and Tomb (Ajmer)

The old tree at the entrance to the tomb is enchanted, it's slit and disfigured with limbs cut off by the keepers of the mosque. No birds sing in its branches because their droppings dirty up the great marble floors.

A smaller tree is green and alive with singing birds: couriers of Heaven and earth for the Islamic faith. Under this tree a saint sat, called the first Sufi in India: Kwajha Mohinddin Cishti, a man who carried water in the bellyskins of goats, a man who poured water on the Hindu totem of *Shiva*—for thirty years the God was pleased.

This man asked *Shiva*, 'I want to be remembered as You are,' and so the water bearer who owned nothing, who loved rich and poor equally, was given a boon and revered as a saint.

Inside the tomb a pink velvet drape swathes the box of one dead, keepers brush the drape over the followers of *Allah*. They kneel and bump foreheads against the flesh coloured coffin, are given rose petals to eat which leave a strange and bitter taste.

Heavy perfume lingers in the tomb, everywhere darkness and ecstasy, the parting of the veil, the slitting of mystery. An ocean of beards, caps, robes and bare feet prostrate before a wall which faces Mecca. Right next to the memory a Muslim is taking money, then he brushes you with a broom to get the sins away.

On the spot of this north/south tomb a man lay down to die, hoping his name would be remembered in the Book of Life; though born a Muslim he had faith in all things, he emanated joy. Now the legend in the mosque repeats itself endlessly—money changers, bigots, women in *purdah* in the swill of desires which grow fat and complacent. Their eyes are dark and fearful as they touch a silver rail, as if that will bring liberation.

The enchanted tree lets everything pass, its black trunk recalls the making of a hero. Silently it stares across at the smaller tree with the nailed on sign: 'Here sat the King of the Poor'.

Chris Wallace-Crabbe

The Secular

However you look at it,
The abundant secular,
How splendid it all appears
Shifting and coruscating
All over and everywhere,
All at once, repeatedly
In little waves of motion
And stubbornly tangible.

Look, I grant all that you say:
Whoever the creator
He brutally botched the job,
But how tough his furniture
Really is made, piece by piece!
I jump on his solid stones
Or dance on these rustling fields
And hear the sap leap in trees
Already marked out for death.

Nor is it Darkness

What will the world do when I am completely gone,
without me to observe it, will it simply blow away
like the milky mist above midwinter football grounds
just after breakfast, your fingers frosty as hell?
What will its beauty add up to if I am plainly not
here to take note, as usual, of heaped cumulo-nimbus
over the plaster pediment-work of 1885
brick terraces, dew on windscreens, the soft
machinery of a turtledove, or the Cootamundra
wattle blushing all over with instantaneous yellow?

The mystical survives. It is not bound by my life,
nor even dependent on quanta.
It merely expands
like the unseen, epiphanic ether

Dusky Tracts

It was the coarsening edge
of the dream attacked me,
thickened with awareness
of enormous dark.

It was the pinegreen melody
that bears no language
began to hurt my knowing,
hard as a wall.

It was blotchy glimpses
of an inkblue deep
as I bobbed all alone
over wavewracked fathoms.

Nothing got said
in that salt orchestration:
strings of the cello
were drawn out of my guts.

Withdrawn from the story-line
was He Who Is,
every image bespeaking raggedly
that which is not.

Sheer Mystery

The uncanny
creeps between ink and paper
chimes through the hairy growing oats and grasses

The uncanny
is deployed exponentially
far beyond extrastellar blackness

It is the dream
capacious enough to contain
all of our littler staccato dreams

It is both
the looping electron and
envisaging the quantum leap

Colourful oceans
roll through it along with
their skirling partner the wind

The uncanny
dissolves our paling fences
between the living and the dead

The uncanny
measures clock time
in a drop of petrified dew

rolling sideways for ever

Maureen Watson

Memo to J.C.

When you were down here JC and walked this earth,
You were a pretty decent sort of bloke,
Although you never owned nothing, but the clothes on your back,
And you were always walking round, broke.
But you could talk to people, and you didn't have to judge,
You didn't mind helping the down and out
But these fellows preaching now in your Holy name,
Just what are they on about?
Didn't you tell these fellows to do other things,
Besides all that preaching and praying?
Well, listen, JC, there's things ought to be said,
And I might as well get on with the saying.
Didn't you tell them 'don't judge your fellow man'
And 'love ye one another'
And 'not put your faith in worldly goods'.
Well, you should see the goods that they got, brother!
They got great big buildings and works of art,
And millions of dollars in real estate,
They got no time to care about human beings,
They forgot what you told 'em, mate;
Things like, 'Whatever ye do to the least of my brothers,
This ye do also unto me'.
Yeah, well these people who are using your good name,
They're abusing it, JC,
But there's people still living the way you lived,
And still copping the hypocrisy, racism and hate,
Getting crucified by the fat cats, too,
But they don't call us religious, mate.
Tho' we got the same basic values that you lived by,
Sharin' and carin' about each other,
And the bread and the wine that you passed around,
Well, we're still doing that, brother.
Yeah, we share our food and drink and shelter,
Our grief, our happiness, our hopes and plans,
But they don't call us 'Followers of Jesus',
They call us black fellas, man.
But if you're still offering your hand in forgiveness
To the one who's done wrong, and is sorry,
I reckon we'll meet up later on,
And I got no cause to worry.

Just don't seem right somehow that all the good you did,
That people preach, not practise, what you said,
I wonder, if it all died with you, that day on the cross,
And if it just never got raised from the dead.

Alan Wearne

St Bartholomew Remembers Jesus Christ as an Athlete

Always in training. Yet helping with his work
was, partly boring, sometimes even nasty.
Still, even when I felt he'd gone too far,
think: here we go again, out came the logic
smooth as a circle, Roman-disciplined.
Brilliant. Yes. Yet never near to God.

Only when he ran.
Only when I saw him striding.
(He'd leap and throw his arms above his head).
It really was a case of 'run with me'.
I did. And often we came down the mountains,
(jogging loosely—never with a cramp).
My running partner—heading for Jerusalem—
appeared as if his feet were next to God.

This too was a feat,—running for a month,
(as rumour had it).
                    Sprinting in the temple
was nothing less than perfect. Tables knocked,
Whips raised and money lost,
He charged them twice.

Of course revenge was needed, and his arms
were raised once more; his feet, however, broken;
sort of enforced retirement. Still,
he made a comeback, to end all comebacks:
                                                            Once
there were ten, and I half-walking, pacing,
(my room mates seated, limbered-up in thought).
We stopped the noise and movement; standing still
I heard the footsteps pounding up the stairs.

Francis Webb

Five Days Old

(For Christopher John)

Christmas is in the air.
You are given into my hands
Out of quietest, loneliest lands.
My trembling is all my prayer.
To blown straw was given
All the fullness of Heaven.

The tiny, not the immense,
Will teach our groping eyes.
So the absorbed skies
Bleed stars of innocence.
So cloud-voice in war and trouble
Is at last Christ in the stable.

Now wonderingly engrossed
In your fearless delicacies,
I am launched upon sacred seas,
Humbly and utterly lost
In the mystery of creation,
Bells, bells of ocean.

Too pure for my tongue to praise,
That sober, exquisite yawn
Or the gradual, generous dawn
At an eyelid, maker of days:
To shrive my thought for perfection
I must breathe old tempests of action

For the snowflake and face of love,
Windfall and word of truth,
Honour close to death.
O eternal truthfulness, Dove,
Tell me what I hold—
Myrrh? Frankincense? Gold?

If this is man, then the danger
And fear are as lights of the inn,
Faint and remote as sin
Out here by the manger.
In the sleeping, weeping weather
We shall all kneel down together.

Poet

I'm from the desert country—O, it's a holy land
With a thousand warm humming stinging virtues.
Masters, my words have edged their way obediently
Through the vast heat and that mystical cold of our evenings.
Many a star, the great lips of wonder drawn out, frozen,
Tempted me, and yet I held my tongue;
But came the long train of camels blowing drowsily:
Words paced, nodding, tinkled through my spirit;
As with the camels, I could never know nor wished to know
Their origin or destiny, for our horizon and the sky
Tremble together in uneasy connubial whiteness.
So my lawless words (I speak figuratively)
Moved the desert, as a train of camels waken
The dozing miles made for them, retreating slowly.

No, my masters, I was never happy there.
It is my first time in this big town of yours.
You are the law, you are the thick grey loam
Of orderly distances, unshakable houses.
Dispel, then, the haze and quandary of my early manhood.
Of course I am still—how shall I put it?—the singer.

And this One you speak of as the enemy of order,
As the wilful floating daze of refractory sunlight:
I do know that we could never exchange words
(But the tinkle and psalm of rubbing harness sometimes
Upon my word and image blowing drowsily . . .?).
Vah! You are the law, my masters, the thick grey loam.
I shall go to the temple with you, take Him in the act;
From the bed of the sick child He comes, from alleyways of the
possessed,
And with this woman He shall speak His public perverseness.

The big stone in my hand will fly shrewdly, I assure you,
As your words, in the house of God.

He stands confronting the woman and death in my hand.
No words between us, I say, for You are the loneliness,
My home, You are the broad light all about me:
You are the train of camels within that light.
Speak up, my masters, quickly, for death hurts my hand.

Cast the first stone. And the grey loam is scattered,
And we slink out one by one. But my narrow clever desert eyes
Peer back over my shoulder. They are strangely together,

A grave broad light in the temple. Breast upon knees,
The woman crouches beside Him: I have seen the sky at midday
Bent earthward. From the two together a train of camels.
She has given her love—but Paradise, what is His love?—
To a hundred of us. Again she will love, may tempt me;
But can ever this stone fly into the face of beauty
While the wind, as His delicate burning finger,
Gives a Word to the sand?

From *Around Costessey*

Good Friday, Norfolk

It is achieved. Over the Place of the Skull
A darkness; and I slouch among our trees,
Tasting green and gold of His darkness. Bid the sun
Withdraw into seismic coma; have interval
Of rootless vacancy enwrap the hill
As vineyard and marshland, mountainside and man:
How the star of the camel-train, the lector seas
Strew antiphon from their substance if but one voice,
The known, dry tongue,
Mould those dismantled words about the Cross.

I blunder through intimate birdcall, marvelling:
Our squadron of tiny charities must go mounting
To feather the sweeps of eclipse, to ripple ponds of sky,
Round-breasted brave nacelle be taking wing.
Birdcall: the round-eyed altimeters sing
And truth veers into joy inevitably
As the coming dewfall adagio earthward, appointing
One rigger cockerel to assemble space.
—Music and light abroad:
The tiny Pilot nested, twisted upon His Cross.

Conscience of the late tractor that straddled cloud,
Was hammer on nailhead, jogs into dozing rumour
Along the sacred stolid flanks of land.
Peter de Draiton, hasten from your shroud
To the church of St Margaret, your beloved bird.
But voices. Eight static centuries pry as wind
Into His darkness; staccato forgiven hammer
Broods in the third hour; and the omnibus
With portly moon-faced grief
Blunders up and down certain hills to know His Cross.

Till I revere our age as the withered Thief
Uttering valid green shoots in his agony.
Till a green visits the earth's unlatched distended frame.
Till the remembered airletter is the brave leaf.
Till speeches, deeds, honour unpondered, burn into life
With the dead man's handclasp, crook a white finger of flame
Upon oil of His darkness that is therapy.
The time is propitious. Dawn in gardener's dress
Stands close to us:
Word of ploughed lands, of sunrise, and a Cross.

Back Street in Calcutta

I have walked among you sorriest skeletons,
Observed that pain is a vacuum—nothing good,
Crevasses in the flesh, emergence of bones;
A little guiltily I shall take my food,
I shall sit playing Bruckner, have his tones
Awaken some pain and anger, then relax,
Or dim sad concepts of order fill my veins:
Of beauty I'll sing to you silent on your backs.

In all your agonies O spare compassion
For me, the well lined and articulate fool
Who knows he tears you, stretched so still, to live.
Tormented flesh that is my flesh, forgive!
And lap around my deathbed like a pool
That starving I may make a true, final confession.

From *Ward Two*

Homosexual

To watch may be deadly. There is no judgment, compulsion,
And the object becomes ourselves. That is the terror:
We have simply ceased, are not dead, and have been
And are; only movement—our movement—is relegated,
Only thought, being—our thought, being—are given
Over; and pray God it be simply given.
So, at this man's ending, which is all a watching,
Let us disentangle the disgust and indifference,
Be all a thin hurried magnanimity:

For that is movement, our movement. Let us study
Popular magazines, digests, psychoanalysts:
For that is thought, being, our thought, our being.

I shall only watch. He is born, seized by joy,
I shall not speak of that joy, seeing it only
As the lighted house, the security, the Beginning.
Unselfconscious as the loveliest of flowers
He grows—and here we enter: the house stands yet,
But the joists winge under our footsteps. Now the God,
The Beginning, the joy, give way to boots and footmarks.
Pale glass faces contorted in hate or merriment
Embody him; and words and arbitrary laws.
He is embodied, he weeps—and all mankind,
Which is the face, the glass even, weeps with him.

The first window broken. Something nameless as yet
Resists embodiment. Something, the perennial rebel,
Will not rest. And this, his grandest element,
Becomes his terror, because of the footsteps, us.

I shall not consider sin beginning, our sin,
The images, furtive actions. All is a secret
But to us all is known as on the day of our birth.
He will differ, must differ among all the pale glass faces,
The single face contorted in hate or merriment.

Comes the day when his mother realises all.
Few questions, and a chaos of silence. Her thin eyes
Are emptied. Doors rattle in the house,
Foundations stagger. The Beginning becomes us;
And he is mulcted of words, remain to him only
The words of sin, escape, which is becoming all of life.
Easier, the talk with his father, rowdy, brief
Thank God, and only the language of the gutter.
He watches the moth pondering the gaslight, love-death,
Offers a wager as to her love or death or both.
His father stops speaking, fingers some papers on the desk.

And now he is here. We had him conveyed to this place
Because our pale glass faces contorted in hate or merriment
Left only sin as flesh, the concrete, the demanded.
He does not speak or hear—perhaps the pox.

But all his compatriots in sin or in other illness
Are flesh, the demanded, silent, watching, not hearing:
It is all he ever sought. Again I am tempted, with the Great,
To see in ugliness and agony a way to God:

Worse, I am tempted to say he has found God
Because we cannot contort our faces in merriment,
And we are one of the Twelve Tribes—he our king.
He has dictated silence, a kind of peace
To all within these four unambiguous walls,
Almost I can say with no answering scuffle of rejection,
He is loving us now, he is loving all.

A Man

He can hardly walk these days, buckling at the knees,
Wrestling with consonants, in raggedy khakis
Faded from ancient solar festivities,
He loiters, shuffles, fingering solid wall:
Away down, the roots, away down,
Who said Let there be light?

The clock in its tower of worked baroque stone
Holds at three o'clock and has always done.
Nothing else shuffles, works, is ended, begun,
There is only the solid air, the solid wall.
Away down, the roots, away down,
Who said Let there be light?

Three weeks under the indigent paid-off clock:
He pulls from his photograph album the heavy chock,
Squats like a king behind a heavy lock,
Niched in and almost part of solid wall.
Away down, the roots, away down,
Who said Let there be light?

Canaries silent as spiders, caged in laws,
Shuffle and teeter, begging a First Cause
That they may tear It open with their claws
And have It hanging in pain from solid wall.
Away down, the roots, away down,
Who said Let there be light?

His King's Cup for swimming, the shimmering girl,
The photogenic light aircraft spin and whirl
Out of the loam, stained by all weathers, hurl
Their petty weight against a solid wall.
Away down, the roots, away down,
Who said Let there be light?

The great goldfish hangs mouthing his glass box
And élite of weeds, like an old cunning fox
Or red-bronze gadfly, hangs in contentment, mocks
All that is cast in air or solid wall.
Away down, the roots, away down,
Who said Let there be light?

But his Cup glitters, the light monoplane bucks
Into the head-wind, girls in panel trucks
Arrive like flowers, and the dry mouth sucks
Deeply, puffs into flesh behind solid wall.
Away down, the roots, away down,
Who said Let there be light?

St Therese and the Child

So too the swaddled mites of mountains taste
Milky droplets of air, old rhymes rolled on the palate
Vowel by vowel westward; and beyond all west
Tablelands of musk are saturating the billet
Where infant or tribe shall toss and turn and drowse,
Blue veins slackening a little: estuaries
Of blood and moon pulse; and this tented rose
Before me while sulky darkness gnaws at our trees.

Your beauty now is all for the King's delight.
Now the last blue lisping bird, Heaven's fondest fool,
Patters after the passionless spindles of light;
Woof and warp wasting, this way and that from the spool.
From some staircase you have let drift one ultimate rose,
Nebulous of skin, but intact as dromedaries
Of the sun at his pride, his prime. To that carnage flows
A little staid creek of cassia, myrrh, aloes.

Madian and Epha, Orange and Lisieux . . .
You are disengaged from the One, and feel your way down
To set amid malnourished arteries of blue
This infant (*mother insane, father unknown*),
Small passions twitching at his face: holy Therese,
A child of two months sleeps here; your sanctuaries
(As if weatherboard breathes) open, open and close
About this so delicate, so inviolable rose.

I am the sun on ticket-of-leave, disconsolate,
Wrack all about me—even the mantled blue
Will not look back. Child, just to consecrate
My last threads of fire to a single blue vein of you . . .
Child at glum sleep's arrest, my parent galaxies
Solicit your calms, times, musing tributaries,
And your Saint—to swing with the tide, to luff and doze
Near your seamless garment of azur, and or, and rose.

Lament for St Maria Goretti

Six o'clock. The virginal belly of a screen
Winces before the blade, the evening wind:
Diluted, a star
Twitches like a puddle on scoured hygienic stone.
All of the documents signed and countersigned
And truce to a cruel war:
Wreckage gesticulates, toothless broken ships,
Meteorite, cherubim, Horseman, in the wash of space
Round the pretty bays of this child's face.

Teresa, it is easier now. But the chloroform
Comes like a stiletto to our gasping void:
Sometimes you look lovely swimming there beside me
While chloroform unravels the holy lines of your face,
I would take you into my hands to remould you, shape you,
But the pain, the pain . . .
See Teresa, my father Luigi is coming
Out of the cemetery (but the chloroform holds) shouldering away
The earth. He touches his little Cross with his lips,
I am crying, and a flight of birds hangs like a rosary,
He is smiling, but the chloroform will dissolve him . . .

Six o'clock. The bells of Nettuno chime
Angelus: Ave to Ave, hand to hand
The buckets of sound are passed in a slow time
Up to a thirsting land.
Again the breeze at the hospital window flutters in lace
Near the thirsting wilderness of this child's face.

Touch me, Teresa: you know you often asked me
Why I was in tears at Mass before the Communion:
I seemed to see Him there, heaving up to Golgotha,
And rising and falling. I stood there mocking Him
Like when I stole two spoonfuls of Angelino's polenta

(You haven't committed a sin, said the kind old priest):
Three times He fell: the last note of the Angelus
Falls with Him—I am falling with Him
—Must I fall with Him into chloroform?
Take up your cross. Touch me. Teresa, quickly . . .

Six o'clock. There may be a moon tonight.
At dead Ferriere twitches the comatose star.
A peasant knows the early mosquito bite
Like a stiletto into his wincing ear.
The suave impersonal light
Trails its skirts over marshland: no mourners here,
And Nothing mourns at Nettuno: feel the embrace
Of Nothing scrambling ashore at this child's face.

Teresa, he's coming: *don't, you will go to Hell* . . .
Teresa, I can still see you: Ferriere is closing in:
The chloroform works at you. Be dainty Corinaldo
Where I was born. I can hardly read or write:
But your breast is our little pet hill, your hair like shadows
Of clouds on our grain, your mouth like a watercourse.
Have you spoken? have words of water been truly uttered
To my thirst—it's this drumming, drumming in my ears.
Teresa, I am going. Teresa, to the last be Corinaldo,
All life writing me on earth:
Let my hands reach you—I can hardly sign my name:
My signature, my scrawl: no wait, Teresa, Teresa . . .

Six o'clock. And the Miserere. Final Grace.
And Death and the Woman, strangely at one, will place
Ambiguous fingers on all of this child's face.

B. R. Whiting

Gandhi, 1946

Did you like him? No. What was it, then?
He had a memorable laugh.
Did you like his disciples? No, not at all,
They had the vestry air.
Yet you remember him? Oh yes, unforgotten
The bright eye and the paper light
Hand on my arm when I walked with him.

Only that? No, time has eaten away
The resentment at a power
That presumed on my arm for its strength,
And now I like to think
How, after I had escorted him to the car,
The Khitmugar came up
To ask permission, if he might touch that arm.

Judith Wright

Eli, Eli

To see them go by drowning in the river—
soldiers and elders drowning in the river,
the pitiful women drowning in the river,
the children's faces staring from the river—
that was his cross, and not the cross they gave him.

To hold the invisible wand, and not to save them—
to know them turned to death, and yet not save them;
only to cry to them and not to save them,
knowing that no one but themselves could save them—
this was the wound, more than the wound they dealt him.

To hold out love and know they would not take it,
to hold out faith and know they dared not take it—
the invisible wand, and none would see or take it,
all he could give, and there was none to take it—
thus they betrayed him, not with the tongue's betrayal.

He watched, and they were drowning in the river;
faces like sodden flowers in the river—
faces of children moving in the river;
and all the while, he knew there was no river.

The Lost Man

To reach the pool you must go through the rain-forest—
through the bewildering midsummer of darkness
lit with ancient fern,
laced with poison and thorn.
You must go by the way he went—the way of the bleeding
hands and feet, the blood on the stones like flowers,
under the hooded flowers
that fall on the stones like blood.

To reach the pool you must go by the black valley
among the crowding columns made of silence,
under the hanging clouds
of leaves and voiceless birds.
To go by the way he went to the voice of the water,
where the priest stinging-tree waits with his whips and fevers
under the hooded flowers
that fall from the trees like blood,

you must forget the song of the gold bird dancing
over tossed light; you must remember nothing
except the drag of darkness
that draws your weakness under.
To go by the way he went you must find beneath you
that last and faceless pool, and fall. And falling
find between breath and death
the sun by which you live.

The Harp and the King

Old king without a throne,
the hollow of despair
behind his obstinate unyielding stare,
knows only, God is gone:
and, fingers clenching on his chair,
feels night and the soul's terror coming on.

Bring me that harp, that singer. Let him sing.
Let something fill the space inside the mind,
that's a dry stream-bed for the flood of fear.
Song's only sound; but it's a lovely sound,
a fountain through the drought. Bring David here,
said the old frightened king.

Sing something. Comfort me.
Make me believe the meaning in the rhyme.
The world's a traitor to the self-betrayed;
but once I thought there was a truth in time,
while now my terror is eternity.
So do not take me outside time.
Make me believe in my mortality,
since that is all I have, the old king said.

I sing the praise of time, the harp replied:
the time of aching drought when the black plain
cannot believe in roots or leaves or rain.
Then lips crack open in the stone-hard peaks;
and rock begins to suffer and to pray
when all that lives has died
and withered in the wind and blown away;
and earth has no more strength to bleed.

I sing the praise of time and of the rain—
the word creation speaks.
Four elements are locked in time;

the sign that makes them fertile is the seed,
and this outlasts all death and springs again,
the running water of the harp-notes cried.

But the old king sighed obstinately,
How can that comfort me?
Night and the terror of the soul come on,
and out of me both water and seed have gone.
What other generations shall I see?
But make me trust my failure and my fall,
said the sad king, since these are now my all.

I sing the praise of time, the harp replied.
In time we fail, alone with hours and tears,
ruin our followers and traduce our cause,
and give our love its last and fatal hurt.
In time we fail and fall.
In time the company even of God withdraws
and we are left with our own murderous heart.

Yet it is time that holds,
somewhere although not now,
the peal of trumpets for us; time that bears,
made fertile even by those tears,
even by this darkness, even by this loss,
incredible redemptions—hours that grow,
as trees grow fruit, in a blind holiness,
the truths unknown, the loves unloved by us.

But the old king turned his head sullenly.
How can that comfort me,
who sees into the heart as deep as God can see?
Love's sown in us; perhaps it flowers; it dies.
I failed my God and I betrayed my love.
Make me believe in treason; that is all I have.

This is the praise of time, the harp cried out—
that we betray all truths that we possess.
Time strips the soul and leaves it comfortless
and sends it thirsty through a bone-white drought.
Time's subtler treacheries teach us to betray.
What else could drive us on our way?
Wounded we cross the desert's emptiness,
and must be false to what would make us whole.
For only change and distance shape for us
some new tremendous symbol for the soul.

Vision

He who once saw that world beyond the world,
so that each tree and building, stone and face
cracked open like a mask before a flame
and showed the tree, the stone, the face behind it—
walked forever with that beatification.
Walking at night, against the blank of darkness,
knew he contained it; touched hand upon brow
and in his gladness cried 'I, even I!'
—knowing the human ends in the divine.

Pride, greed, and ignorance—that world's three veils—
through them he walked and saw what lay beyond;
saw what the human eye was meant to see:
and watched the greedy and the stupid fumble
in a blind fear with intellect and pride—
those blades that cut the ignorant hands that hold them.
So he was sad for victim and oppressor,
for crying child and brute with the slack mouth,
for schemer, clod and safe respectable man
and all who had not seen what he had seen.

And yet these, too, moved in that second world
and stood up real behind the masks of hatred.
The very wound and weapon bled and glittered
as though both steel and flesh were made of light
and men the instruments in some high battle
where God incomprehensibly warred on God.

Wherefore he closed his eyes and hands, and prayed
vision and action know their proper limits,
and knowledge teach him more humility.

Prayer

Let love not fall from me though I must grow old.
To see the words fade on the fading page,
to feel the skin numbing in fold on fold,
the mind and the heart forgetting their holy rage—

oh no, let me run, till the wind's agues blow
my cinders red again—let me tilt and drain
the last drop of my life before I go.
Let the earth's choirs and messengers not sing in vain.

While every flower swings open its eternal door
and every fruit encloses its timeless seed
let me not watch in spite, caring no more,
but let my heart's old pain tear me until I bleed.

Out in the dark, I know, sing a thousand voices;
and the owl, the poet's bird, and the saint's white moth
blunder against my window, the frog in the rain rejoices.
I pledge to the night and day my life's whole truth.

And you, who speak in me when I speak well,
withdraw not your grace, leave me not dry and cold.
I have praised you in the pain of love, I would praise you still
in the slowing of the blood, the time when I grow old.

For One Dying

Come now; the angel leads.
All human lives betray,
all human love erodes
under time's laser ray;

the innocent animals
within us and without
die in corrupted hells
made out of human thought.

Green places and pure springs
are poisoned and laid bare—
even the hawk's high wings
ride on a fatal air.

But come; the angel calls.
Deep in the dreamer's cave
the one pure source upwells
its single luminous wave;

and there, Recorder, Seer,
you wait within your cell.
I bring, in love and fear,
the world I know too well

into your hands. Receive
these fractured days I yield.
Renew the life we grieve
by day to know and hold.

Renew the central dream
in blazing purity,
and let my rags confirm
and robe eternity.

For still the angel leads.
Ruined yet pure we go
with all our days and deeds
into that flame, that snow.

Eight-panel Screen

Here the Sage is setting out.

A simple garment, cloth of blue,
is gathered in his girdle. Bare
head, rope sandals; seven lines
circumscribe him; that will do.
Now the world stands round about:
a path, a tree, a peak in air,
one narrow bridge beneath the pines.

Here's the Boy, three steps behind.

A cooking-pot, a sag-backed horse,
and his master's steps to tread
with a bundle on his back,
a tuft of hair, a stick of course,
rounded face still undefined.
As the Sage goes on ahead
the horse's rope takes up its slack.

Now the path begins to climb.

But the Sage still knows the Way,
sets his profile like a crag
or an eagle; meets the storm,
never waiting to survey
World in a moment's breathing-time.
On go Boy and stolid Nag.
Tao knows neither cold nor warm.

Now the path goes down the hill.

Steadily the Sage descends;
Boy and Horse go patter-clop
past the charcoal-burner's hut
where the crooked pine-tree stands.
On the Sage goes striding still.
Droops the Horse's underlip?
Does Boy falter in his trot?

Now they skirt the mountain-brook.

Past the fishers with their rods,
past the children in their game,
past the village with its smoke
and the ploughman in his clods;
up again the path goes—look!
Boy is dragging, so is Moke,
but to Sage it's all the same.

Up—and this time higher yet.

How, Boy wonders, be a Sage?
How ignore such aching feet
only thinking of the Way?
Wisdom seems to come with age—
if it's wisdom to forget
Stomach's groaning yawn for meat
and keep striding on all day.

Round and round the stairways wind.

Cloud and pine-tree, rock and snows,
surround the Sage's sinewy lope.
Muscles strung to meet the steep,
how his one blue garment blows!
Boy is rather far behind;
Horse is leaning on his rope;
Even Sun sinks down to sleep.

Look! The rest-house, there at last.

Sage sits down to meditate,
Moon accosts the last of day.
Boy brings water, stumbling now;
sees his face there fluctuate—
not so round! More sternly cast!
Patience and the endless Way,
these refine us. *That* is Tao.

Grace

Living is dailiness, a simple bread
that's worth the eating. But I have known a wine,
a drunkenness that can't be spoken or sung
without betraying it. Far past Yours or Mine,
even past Ours, it has nothing at all to say;
it slants a sudden laser through common day.

It seems to have nothing to do with things at all,
requires another element or dimension.
Not contemplation brings it; it merely happens,
past expectation and beyond intention;
takes over the depth of flesh, the inward eye,
is there, then vanishes. Does not live or die,

because it occurs beyond the here and now,
positives, negatives, what we hope and are.
Not even being in love, or making love,
brings it. It plunges a sword from a dark star.

Maybe there was once a word for it. Call it grace.
I have seen it, once or twice, through a human face.

Patterns

'Brighter than a thousand suns'—that blinding glare
circled the world and settled in our bones.

Human eyes impose a human pattern,
decipher constellations against featureless dark.

All's fire, said Heraclitus; measures of it
kindle as others fade. All changes yet all's one.

We are born of ethereal fire and we return there.
Understand the Logos; reconcile opposing principles.

Perhaps the dark itself is the source of meaning,
the fires of the galaxy its visible destruction.

Round earth's circumference and atmosphere
bombs and warheads crouch waiting their time.

Strontium in the bones (the mass-number of 90)
is said to be 'a good conductor of electricity'.

Well, Greek, we have not found the road to virtue.
I shiver by the fire this winter day.

The play of opposites, their interpenetration—
there's the reality, the fission and the fusion.

Impossible to choose between absolutes, ultimates.
Pure light, pure lightlessness cannot be perceived.

'Twisted are the hearts of men—dark powers possess them.
Burn the distant evildoer, the unseen sinner.'

That prayer to Agni, fire-god, cannot be prayed.
We are all of us born of fire, possessed by darkness.

Yirrkalla People of Arnhem Land

From *The Djanggawul Song Cycle*

Song 20

What is that, *waridj*? Carefully sway our buttocks, Bralbral, with dragging paddle.
This is sea-weed floating, a mass of sea-weed floating . . .
Waridj, it is a stranger from Bralgu, this floating weed, with fragments dislodging and drifting away.
It drifts into rows spreading out with the waves. Blown by the wind, it drifts backwards and forwards—spreading across the water.
Roar of the sea as it carries the weed on the incoming tide, and the spray of the surf!
We saw it through the shine of the Morning Star, there on our way to Port Bradshaw from Bralgu.
Let us go fast, *waridj* Djanggawul. Moving our arms quickly and swaying our buttocks, we lift up the paddles.
Far we have paddled. Is that spray, rising before us?
Is that the sea smell we follow? Is that the Baijini sound we hear?
Let us paddle fast, *waridj,* quickly, moving our buttocks, as the waves rush in, the high tide breaks on the beach . . .
Carefully, *waridj.* Djanggawul, moving his buttocks, lifts up his narrow paddle.

Song 21

What do we see, *waridj,* as we look back?
Paddling, we see the shine of the Morning Star.
Yes, *waridj* Bralbral, paddling we see it close to us. The Morning Star sends out its rays as it rises near us:
It skims the water, shining across the sea, the Bralgu Star, its rays shining near us.
It skims the sea, from Bralgu, shining upon us, on the end of its string, attached to a young sapling.
Another Star, *waridj,* a feathered ball held by the Spirits . . . Close is the Morning Star . . .
It shines near, as we turn to see it. Oh, Morning Star, Oh pole and strings . . .
The Star and its rays rise gradually for us, *waridj* Miralaidj; we rest our paddles, dragging them through the sea.

See the shine from the disc of the Star, close to us, *waridj*.
The Bralgu Spirits are dancing, sending the Star . . .
Rain people, the stamping sound of their dancing . . .! Dust rises, *waridj*, from under their dancing feet . . .!
There from the fresh water place, from the Spirit country.
The shine falls on us, from Bralgu, covering us with its shine . . . Close, it shines across to the mainland, to Port Bradshaw.
The morning Star shines, bringing the dawn . . . putting an end to the night . . .
Close to us, ending the darkness, bringing the dawn . . .
It pierces the darkness, that Star, sent by the Spirits. The sound of their dancing!
The morning Star shines from Bralgu, shines like the *mauwulan* pole . . .
It rises from the sandhills at Bralgu, where the Spirits dance, *waridj*, for us.
We move fast with the narrow paddle; we hasten, moving our wrists, as the tide roars in.
For the daylight is on us, the dawn, before the cry of the morning pigeon. The Star still shines for us, *waridj* Djanggawul.
Our hips and buttocks are swaying. Let us go carefully, because of the water,
It is rising for us, and roaring, with foam and spray of the surf.
We push the water along with our canoe, our paddles swishing.
Waves come up, with the rising tide. Spray comes from our swishing paddles. From us the tide is flowing.
Is that the mainland we are approaching from Bralgu?
Is that the mainland, Port Bradshaw, we are approaching . . .?
The smell of the sea! *Waridj*, we paddle fast, quickly moving our wrists.
We lift up the narrow paddles—*waridj* Bralbral, our paddles . . .
We lift up the big paddles, moving fast, swishing the paddles . . .
Our paddles! We drag them along, the flat and the narrow paddles . . .

Song 22

What is that, *waridj*, in front of us? We rest, dragging our paddles.
There ahead of us, *waridj* Bralbral.
That is the morning pigeon. Darkness goes, with its cry!
We saw darkness only towards the west . . .
Bird that 'twists' it tongue as it whistles! Daylight comes with its cry.
We saw the calmness of dawn (no sound but the pigeon's cry).

It ruffles itself, *waridj,* crying, and shaking its feathers. Its cry goes out to its nestlings.
The small clear cry of that pigeon and of its nestlings . . .
Talking fast from their nest . . . The pigeon flies down from the smooth inside of the nest. Their cries reach us, 'twisting' their tongues . . .
We saw the Morning Star ending the darkness. Then, the cry of the pigeon! Cries like the speech of different linguistic groups, like the Madarlpa dialect!
The bird talks fast, a sound like the Dalwongu dialect . . .
It flew across from its nest in the 'arm-band' bushes . . .
It teaches that cry to its young: the nestlings ruffle themselves.
Teaching that cry! It calls from the limbs near its nest. They are 'twisting' their tongues . . .
We saw it, flying down, as the darkness was clearing, 'twisting' its tongue and whistling.

Song 23

We paddle fast, moving our buttocks, *waridj.* We lift up the narrow paddle . . .
We paddle fast, with the narrow paddle . . . We speed along, *waridj* Djanggawul . . .
Wei! What is that, *waridj*? We rest our paddles, Bralbral, and drag them.
That, *waridj,* is the long drawn cry of the black bird.
Crying out, for it saw the darkness clearing . . .
What is that? The sound of the black bird, long drawn as the darkness clears.
It saw the coming dawn, and the calmness.
What is that, *waridj*? A black bird crying, at dawn, and the cry of the nestlings!
Crying with mouths agape, as they hear the roar of the waves, and smell the sea,
As they hear the noise of the water, the foam and the spray.
'I (says the bird) heard the noise of the water rising, the incoming tide . . .
I draw out my sound, it reaches out to Port Bradshaw . . .
A long cry, moving my head, as I see the water rising, and see the spray . . .
Nesting among the limbs of the "coffee" tree, of the *djuda rangga* . . .'
It grasps the tree with its claws, crying from that sacred tree,
Sending out its cry, as the big waves rise and spray.

The sea roars as the bird cries, and the sound reaches up to the clouds,
A long-drawn sound entering the banking clouds, the cloud-flecked sky.
A long cry, as it saw the tide roaring in, with its spray
'I saw the spray. I cry out as the waves come splashing together.
I saw the water, and saw them paddling fast. The smell of the sea!
I saw the tide coming up. The Baijini sound, the smell of the sea!
I saw the water, roaring, as the waves came in at Port Bradshaw, coming from Bralgu.'
Foam coming up, and spreading. The bird cried as it looked at the water.
The crest of the surf shines in the light of the Morning Star. The sea roars, and the waves are splashing together.
The bird saw the darkness clearing, uttering its long cry as the dawn came,
The cry in the stillness, as the darkness cleared (with the light) from Bralgu!

Song 24

What is that, *waridj* Bralbral? We rest our paddles at the cry, and drag them along.
We look round to see the cry. It comes from Bralgu. It's close to us!
It is coming up, moving along. The sun, with its *mauwulan rangga* emblem!
As it rises, its rays warm the Djanggawul's backs. It comes rising, close to them.
The sun, with rays emerging before its disc! Close, it shines on the water, warming our backs, *waridj* Djanggawul.
Close it is rising, from the sand, from the sea into the sky.
Hot sun, burning our backs, its rays leading back to Bralgu!
Rising sun, reflected in the sea! Sun for us, with its heat!
Rays of the sun emerging, *waridj,* leaving its home beneath the water!
Sun coming up close, with spreading rays!
For us, *waridj,* it leaves its home under the sea.
Its rays touch us like hands; their reflections shine in the water.
We go fast. Paddling along we see the rays touching us, so that the sweat comes out.
The sun comes up for us, leaving its home; the glare is hurting our eyes.
It comes closer, rising above the sea, burning our backs.

The smell of the sea! It leaves its home in the water, and warms us . . .
We paddle fast, to Port Bradshaw, to the Place of the Sun, to where the Baijini are.
Reflections shine in the sea. The rays warm us, on our way to Port Bradshaw.
We lift up our paddles, *waridj*, we go fast but carefully, moving our arms and our buttocks.
The sun reaches us, with its rays. Its red glow, *waridj*, for us, from the sacred parakeet feathers!
We rest our paddles. It is for us! We paddle fast, *waridj*, and carefully.
Our buttocks are moving. It is for us, *waridj* Djanggawul! We paddle fast!
Our hips sway as we paddle along, lifting the paddles,
On our way to the Place of the Sun, at Port Bradshaw.
The sun's rays touch us, warming our backs . . .
Rays like the parakeet-feathered string of our *rangga*! Feathered string like our child! Red glow of the sun.
Rays warming our backs, like feathered string! Like feathered *rangga*!
That sun rises above us, burning our backs, going to the Place of the Sun.
It burns our backs, and it shines on the water at Lilildjang.
Its warm rays touch us, stretching to Arnhem Bay . . .
It leaves its home in the water, and rises, burning.
Warm rays touching our backs, touching our *rangga*! Shining, and making the mainland clear to see!
For us it shines on the sacred sandhills at Port Bradshaw . . .
Stretching out its rays, warming our backs, illuminating the water-holes at Arnhem Bay . . .
That sun, sending out its rays to Elcho Island,
Burning our backs, from the red ochre there . . . at Elcho Island . . .
That sun sends out its rays to shine on the sea, *waridj*, on the mainland near Milingimbi . . .,
Warming our backs, as it reaches west to the wide Barara country.

Song 25

We go fast, moving our wrists, and lifting our paddles.
We paddle along, *waridj* Bralbral, as the sun's rays touch us, and touch our paddles.

Our buttocks and hips sway, as we paddle along:
What is that? What is that cry?
Flying foxes, suspended there in the tree, *waridj* Djanggawul, crying out from the tree.
We saw the sacred tree, the sacred *rangga*. Flying foxes, in the sacred tree!
Little flying foxes crying, as they hold that sacred tree . . .!
They cry from their home, where they hang among the branches of the sacred tree,
Cry from the topmost branches, from all the branches of the tree . . .
They cry from their home: crying, *waridj,* from the sacred Leichhardt fig tree.

Song 26

We go fast, lifting our paddles, moving our wrists . . . we are close to our country.
We go fast, moving our buttocks; for we hear the roar of the sea, the spray of the surf,
Coming up from our paddling! Foam from our paddling!
Wai! *waridj* Bralbral, carefully drag the paddles, for they are splashing up foam.
What has happened? Sea smell, and splashing!
What, *waridj,* is that? Wai! Our feathered arm-bands!
What has happened? They are wet from that foam.
We are all getting wet! What shall we do? Throw away the wet *rangga*?
They are wet from that foam, our *rangga, waridj* Bralbral: wet from lying in water! We are nearing Port Bradshaw.
Shall we, *waridj,* undo it, then throw it away? No, let the sacred mat with the *rangga* sink down, near the Place of the Sun.
We shall not untie its 'mouth' . . . Let it sink down into the sea, outside Port Bradshaw . . .
Sacred invocations to the *rangga* and *ngainmara* . . .! We invoke the *bugali* as it sinks . . ., calling the sacred names.
They sink down, making a noise as the sea covers them.

Song 27

Let us rest on our paddles, *waridj,* for I (Bralbral) am tired.
Stop paddling, *waridj.*
What is happening there, *waridj*? (Bralbral says) My body aches with tiredness.

Tired, because you are worrying (about the mat containing the *rangga*).
I am worrying (says the Djanggawul Brother) about the *rangga*. (Why didn't we open the mat?)
I am just tired! That was why, *waridj*, we threw away the sacred *rangga* and mat.
We are coming close to the mainland: our journey, our paddling, is done.
We land on the beach at Port Bradshaw.
That is our country! We plant our *mauwulan* here.
We have arrived, oh *waridj*!

TRANSLATED BY RONALD M. BERNDT

Fay Zwicky

From *Ark Voices*

Lemur

This powerful tail this tiny brain
Would make of man an ass:

Because of such as me the earth is riven.
Yet I suit your plan and am

Forgiven for it. Lord, you
Surge me clear of pain:

Nor past nor future, duty nor
Regret are mine.

Today today today only
Today I swing upon my ring—

Tailed rung from sleep to sleep to
Hunger   Play   Sleep   to

Leap through blackest night.
A state of grace.

Laved in eternal rhythms
My ailing howls kill time:

We eat and die.

Your eye burns a dark
Angelic arc into my frightened fur.

The rain will wash it clean
Away—before me the flood—

Io Lemuria! Wandering spirit of the dead
Voracious once a year revisit

Those I loved who feed me quick,
Thud wide the door, and shaking pray:

'Manes exite Paterni!'
Leave! Ghosts of our fathers!

Grafted with their pain I go
Only to return

Sir, I fear my part in this haunting

Whale Psalm

I steer the chastened furrows
with my tail
coil filamented upwards   lift   thrash
down to
crash the
heaving waste
behind.

My captors close
upon me, sir, I call—

                                        Thew and sinew
peak and plunge: then softly softly
stealthy roll and glide, recoil to
coil again

                                        lift in subtle curvature
plunge downward:
my ponderous flukes subdue
the darkening flood.

                                        O sir, you thus
prepared me, thus I churned your path
chanted your praise: my being
spoke your wonder.

                                        Unmoored from innocence
from your sight cast, today I range
hell's belly.

Earth's nets tighten:
men forsake their mercy, shroud me dumb
who have so loved the habitation of
your waters.

Rein me from darkness now as once
you ransomed Nineveh lest
fishers mourn, nets languish
on the blackening sea.

ACKNOWLEDGMENTS

We wish to thank the copyright holders for permission to reproduce the following material:

Robert Adamson: 'The River' from *Selected Poems 1970–1989*, reprinted by permission of University of Queensland Press, 1990. **Anbarra People of Arnhem Land:** 'White Cockatoo' (performed by Frank Malkorda, 1978; translated by Margaret Clunies Ross), 'Wild Honey and Hollow Tree' (performed by Frank Gurrmanamana, 1975; translated by Margaret Clunies Ross), 'Crow' (performed by Frank Malkorda, 1982; translated by Margaret Clunies Ross) from *The Honey-ant Men's Love Song*, Eds R. M. W. Dixon and Martin Duwell, published by University of Queensland Press, 1990. **Dorothy Auchterlonie:** 'Resurrection' from *The Dolphin*, ANU Press, 1967. **'William Baylebridge'** (William Blocksidge): 'Life's Testament: II; III', 'Salvation', 'Deity' from *The Collected Works of William Baylebridge*, vol. 1, Angus & Robertson, a division of HarperCollins Publishers, 1961. **Bruce Beaver:** 'Lauds and Plaints: III' from *Lauds and Plaints, Poems 1968–1972*, South Head Press, 1974, reproduced by permission of the poet. **Judith Beveridge:** 'The Herons', 'The Dispossessed Angels', 'Performing Angels' from *The Domesticity of Giraffes*, Black Lightning Press, 1987. **John Blight:** 'It', 'Ant, Fish and Angel', 'Deities' from *Selected Poems 1939–1990*, University of Queensland Press, 1992. **Francis Brabazon:** 'Well have you called yourself the Ocean of Mercy—' from *In Dust I Sing*, Beguine Library, 1974. **Christopher Brennan:** 'Farewell, the pleasant harbourage of Faith', 'Towards the Source: VII' from *Selected Poems*, Angus & Robertson, a division of HarperCollins Publishers, 1992. **Vincent Buckley:** 'Before Pentecost', 'Song for Resurrection Day' from *Masters in Israel*, Angus & Robertson, a division of HarperCollins Publishers, 1961; 'Puritan Poet Reel', 'Places', 'Eleven Political Poems: No new thing; Day with its dry persistence' from *Selected Poems*, Angus & Robertson, a division of Harper-Collins Publishers, 1981. **Ada Cambridge:** 'Vows' from *The Hand in the Dark and Other Poems*, Heinemann, 1913. **David Campbell:** 'Far Other Worlds', 'Speak with the Sun', 'The Miracle of Mullion Hill', 'Fisherman's Song', 'Among the Farms', 'Cocky's Calendar: Prayer for Rain', 'Trawlers', 'A Yellow Rose' from *Collected Poems*, Angus & Robertson, a division of HarperCollins Publishers, 1981. **Gary Catalano:** 'The Jews Speak in Heaven', 'A Dream of Hell', 'Heaven of Rags' from the author's *Selected Poems, 1973–1992*, University of Queensland Press, 1993; 'Equation' from *Fresh Linen*, University of Queensland Press, 1988; 'Hole' from *The Empire of Grass*, University of Queensland Press, 1991. **Alison Clark:** 'Credo', 'Ananke', 'Respecting the Mysteries' from *Ananke*, published in *Scripsi*, 1987, reproduced by permission of the poet; 'Gardening' reproduced by permission of the poet. **David Curzon:** 'Psalm 1', 'Proverbs 6:6' from *Midrashim*, Cross-Cultural Communications, 1991, © David Curzon, reproduced by permission of the poet and publisher. **Bruce Dawe:** 'Happiness is the art of being broken', 'And a Good Friday Was Had by

All', 'Bring out Your Christs', 'A Week's Grace', 'At Mass', 'The Christ of the Abyss' from *Sometimes Gladness: Collected Poems, 1954–1978*, Longman Cheshire Pty Ltd, 1988. **Rosemary Dobson:** 'Two Visions', 'The Missal', 'Callers at the House', 'Being Called For', 'The Almond-tree in the King James Version' from *Collected Poems*, Angus & Robertson, a division of HarperCollins Publishers, 1991; 'The Apparition' from *Untold Lives*, Brindabella Press, 1992. **Michael Dransfield:** 'Geography: III', 'Morning Prayer' from *Collected Poems*, edited by Rodney Hall, University of Queensland Press, 1987. **Geoffrey Dutton:** 'Twelve Sheep' from *New Poems to 1972*, Australian Letters, 1972, reproduced by permission of the poet. **'E'** (Mary E. Fullerton): 'Earth', 'Impregnable', 'Windows' from *The Wonder and the Apple*, Angus & Robertson, a division of HarperCollins Publishers, 1946. **Stephen Edgar:** 'Tenebrae' from *Ancient Music*, Angus & Robertson, a division of HarperCollins Publishers, 1988; 'Dead Souls' from *Corrupted Treasures*, William Heinemann, in press, reproduced by permission of the poet. **Diane Fahey:** 'Millipede at an Ashram' from *Mayflies in Amber*, Angus & Robertson, a division of HarperCollins Publishers, 1993. **Robert D. FitzGerald:** 'The Greater Apollo: IV', 'Revelation', 'Eleven Compositions: Roadside: VI', 'Insight: Six Versions: Creak of the Crow' from *Forty Years' Poems*, Angus & Robertson, a division of HarperCollins Publishers, 1965. **John Forbes:** 'Ode to Doubt' from *New and Selected Poems*, Angus & Robertson, a division of HarperCollins Publishers, 1992. **John Foulcher:** 'Wars of Imperialism', 'Reading Josephus' from *New and Selected Poems*, Angus & Robertson, a division of HarperCollins Publishers, 1993. **Barbara Giles:** 'Beauty for Ashes' from *The Hag in the Mirror: New and Selected Poems*, Pariah Press, 1989. **James Gleeson:** 'Drill of Central Thunder Notes: the city expects Christ:' from *Selected Poems*, Angus & Robertson, a division of HarperCollins Publishers, 1993. **Peter Goldsworthy:** 'Mass for the Middle-Aged: Libera Me' from *This Goes with That: Selected Poems 1970–1990*, Angus & Robertson, a division of HarperCollins Publishers, 1991; 'A Brief Introduction to Philosophy: What Comes Next?' from *Little Deaths*, Angus & Robertson, a division of HarperCollins Publishers, 1993, reproduced by permission of the poet. **Alan Gould:** 'The Henty River' from *Selected Poems*, Angus & Robertson, a division of HarperCollins Publishers, 1992, © Alan Gould, c/- Margaret Connolly & Associates Pty Ltd. **Robert Gray:** 'Dharma Vehicle' from *Selected Poems* (revised edition), Angus & Robertson, a division of HarperCollins Publishers, 1985. **Lesbia Harford:** 'Summer Lightning', 'Buddha in the Workroom', 'A Deity', 'I am no mystic', 'A Prayer to Saint Rosa' from *The Poems of Lesbia Harford*, Angus & Robertson, a division of HarperCollins Publishers. **Charles Harpur:** 'The Silence of Faith' from *Selected Poetry and Prose*, Penguin Books Australia, 1986. **Robert Harris:** 'Ray', 'The Call', 'The Cloud Passes Over', 'Isaiah by Kerosene Lantern Light', 'The Eagle' from *The Cloud Passes Over*, Angus & Robertson, a division of HarperCollins Publishers, 1986. **Kevin Hart:** 'Master of Energy and Silence', 'The Stone's Prayer', 'Approaching Sleep', 'Facing the Pacific at Night', 'The Gift' from *New and Selected Poems*, Angus & Robertson, a division of HarperCollins Publishers, in press. **William Hart-Smith:** 'Christopher Columbus: Psalm for Himself', 'Negative', 'Aaron's Rod', 'Ambrosia' from *Selected Poems 1936–1984*, Angus & Robertson, a division of HarperCollins

Publishers, 1985. **Gwen Harwood:** 'A Case', 'Home of Mercy', 'Midnight Mass, Janitzio', 'Revival Rally', 'The Wasps' from *Selected Poems*, Angus & Robertson, a division of HarperCollins Publishers, 1990; 'Class of 1927: Religious Instruction', 'Resurrection' from *Bone Scan*, Angus & Robertson, a division of HarperCollins Publishers, 1988; 'Night Thoughts' from *Night Thoughts*, National Library of Australia, 1990, reproduced by permission of the poet. **Kris Hemensley:** 'the white daisies' from Site, work-in-progress, 1986–94, reproduced by permission of the poet. **A. D. Hope:** 'Easter Hymn', 'An Epistle from Holofernes', 'A Bidding Grace', 'A Letter from Rome', 'Faustus' from *Collected Poems 1930–1970*, Angus & Robertson, a division of HarperCollins Publishers, 1972; 'The Mystic Marriage of St Catherine of Alexandria' from *A Late Picking*, Angus & Robertson, a division of HarperCollins Publishers, 1975; 'Visitant' from *Orpheus*, Angus & Robertson, a division of HarperCollins Publishers, 1991. **Kate Jennings:** 'Saint Munditia' from *Cats, Dogs and Pitchforks*, William Heinemann, 1993. **Evan Jones:** 'Servetus: October 27th, 1553' from *Inside the Whale*, Longman Australia Pty Ltd, 1960. **Antigone Kefala:** 'Alter Ego', 'Worship' from *Absence: New and Selected Poems*, Hale & Iremonger, 1992. **Henry Kendall:** 'Dedication: To a Mountain' from *The Poetical Works of Henry Kendall*, Libraries Board of South Australia, 1966. **Peter Kocan:** 'Cathedral Service' from *Standing with Friends*, William Heinemann, 1992, reproduced by permission of the poet. **Anthony Lawrence:** 'God is with me as I write this down' from *Dreaming in Stone*, Angus & Robertson, a division of HarperCollins Publishers, 1989. **Geoffrey Lehmann:** 'Spring Forest: Mother Church; Witnesses' from *Spring Forest*, Angus & Robertson, a division of HarperCollins Publishers, 1992. **James McAuley:** 'To the Holy Spirit', 'New Guinea', 'Pietà', 'Father, Mother, Son', 'One Tuesday in Summer' from *Collected Poems 1936–1970*, Angus & Robertson, a division of HarperCollins Publishers, 1992; 'Music Late at Night' from *Music Late at Night*, Angus & Robertson, a division of HarperCollins Publishers, 1976. **Hugh McCrae:** 'Down the Dim Years' from *The Best Poems of Hugh McCrae*, Angus & Robertson, a division of HarperCollins Publishers, 1961. **Nan McDonald:** 'Sunday Evening' from *Pacific Sea*, Angus & Robertson, a division of HarperCollins Publishers, 1947; 'The Barren Ground', 'The Return', 'The Last Mile' from *Selected Poems*, Angus & Robertson, a division of HarperCollins Publishers, 1969. **Kenneth Mackenzie:** 'How near, o god . . .', 'No body . . .' from *The Poems of Kenneth Mackenzie*, Angus & Robertson, a division of HarperCollins Publishers, 1972. **David Malouf:** 'Metamorphoses' from *Selected Poems*, Angus & Robertson, a division of HarperCollins Publishers, 1991. **J. S. Manifold:** 'Red Rosary: Death of Stalin', 'Astronauts', 'Nativity' from *Collected Verse*, University of Queensland Press, 1978. **Leonard Mann:** 'Vision' from *The Delectable Mountains and Other Poems*, Angus & Robertson, a division of HarperCollins Publishers, 1944; 'To God' from *Elegiac and Other Poems*, Longman Cheshire Pty Ltd, 1957. **Billy Marshall-Stoneking:** 'Sky' from *Singing the Snake*, Angus & Robertson, a division of HarperCollins Publishers, 1990. **Philip Martin:** 'An English Martyr', 'A Sacred Way' from *New and Selected Poems*, Longman Cheshire Pty Ltd, 1988. **Mudrooroo:** 'The Song Circle of Jacky: Song 22' from *The Garden of Gethsemane*, reproduced with the kind permission of Hyland House, 1991. **Les Murray:** 'An Absolutely Ordinary

Rainbow', 'The Future', 'The Chimes of Neverwhere', 'At Min-Min Camp' from *Collected Poems*, Angus & Robertson, a division of HarperCollins Publishers, 1991, reproduced by permission of Les Murray, c/- Margaret Connolly & Associates Pty Ltd, PO Box 48, Paddington, NSW, 2021; 'The Barranong Angel Case' from *The Weatherboard Cathedral*, Angus & Robertson, a division of HarperCollins Publishers, 1969, reproduced by permission of Les Murray, c/- Margaret Connolly & Associates Pty Ltd, PO Box 48, Paddington, NSW, 2021. **John Shaw Neilson:** 'Love is a Fire', 'The Worshipper', 'Surely God was a Lover', 'He was the Christ', 'Song for a Sinner', 'Schoolgirls Hastening', 'To the Untuneful Dark', 'The Crane is My Neighbour', 'The Vassal' from *Selected Poems*, Angus & Robertson, a division of HarperCollins Publishers, 1992. **Geoff Page:** 'Country Nun' from *Smalltown Memorials*, University of Queensland Press, 1975; 'Broken Ballad', 'My Mother's God' from *Selected Poems*, Angus & Robertson, a division of HarperCollins Publishers, 1991; 'Curtains, Death and Me' from *Human Interest*, Heinemann, 1994, reproduced by permission of the poet. **Peter Porter:** 'Who Gets the Pope's Nose?', 'The Old Enemy', 'Looking at a Melozzo da Forlì', 'An Angel in Blythburgh Church', 'The Unlucky Christ' from *Collected Poems*, Oxford University Press, 1983. **Craig Powell:** 'The Water Carrier' from *Minga Street: New and Selected Poems*, Hale & Iremonger, 1993. **Jennifer Rankin:** 'After Meditation' from *Collected Poems*, University of Queensland Press, 1990. **Elizabeth Riddell:** 'The Memory', 'Ecce Homo' from *Selected Poems*, Angus & Robertson, a division of HarperCollins Publishers, 1992. **Nigel Roberts:** 'Reward / for a missing deity' from *In Casablanca for the Waters*, Wild & Woolley, 1977. **Roland Robinson:** 'Invocation' from *Selected Poems*, Kardoorair Press, 1983; 'The Curlew', 'The Sermon of the Birds' from *Selected Poems*, Angus & Robertson, a division of HarperCollins Publishers, 1989. **David Rowbotham:** 'God of the Cup and Planet' from *The Makers of the Ark*, Angus & Robertson, a division of HarperCollins Publishers, 1970, reproduced by permission of the poet. **Noel Rowe:** 'Magnificat: I', 'Bangkok: III' originally published in *Quadrant*, reproduced by permission of the poet. **Philip Salom:** 'Inquiry of the Spirit Body' from *Barbecue of the Primitives*, University of Queensland Press, 1989. **John A. Scott:** 'Four Sonnets: Theatre of the Dead Starling' from *The Barbarous Sideshow*, Makar Press, 1976, reproduced by permission of the poet; 'Limbo', 'Waltz' from *The Quarrel With Ourselves*, Rigmarole Books, 1984, reproduced by permission of the poet. **Thomas Shapcott:** 'Singing in Prison', 'Portrait of Saul' from *Thomas Shapcott Selected Poems 1956–1988*, University of Queensland Press, 1989. **Jemal Sharah:** 'Reliquaries', 'Miserere', poet holds copyright; 'Revelation', published in *Quadrant*, poet holds copyright. **R. A. Simpson:** 'Lachryma Christi', 'Tunnels' from *Selected Poems*, University of Queensland Press, 1981, reproduced by permission of the poet; 'The Iconoclast' from *Diver*, University of Queensland Press, 1972, reproduced by permission of the poet. **Alex Skovron:** 'A Life: The Moth', 'Sleeve Notes: Strings' from *Sleeve Notes*, Hale & Iremonger, 1992. **Peter Skrzynecki:** 'Pietà' from *Head-Waters*, Lyre-Bird Writers, 1972. **Vivian Smith:** 'The Other Meaning', 'Return of the Prodigal Son', 'The Traveller Returns', 'From Korea' from *Selected Poems*, Angus & Robertson, a division of HarperCollins Publishers, 1985. **Vyvien Starbuck:** 'Holy Thursday', published in *Poetry Australia*,

1983, reproduced by permission of the poet; 'Crucifixion', published in *Public Works: III*, ANU Press, reproduced by permission of the poet. **Peter Steele:** 'A.D. 33', 'Matins' from *Word from Lilliput*, The Hawthorn Press, 1973; 'To Thomas More', 'Covenant: I' from *Marching on Paradise*, Longman Cheshire Pty Ltd, 1984; 'Of God and Despair', published in *The Age*, reproduced by permission of the poet. **Harold Stewart:** 'Lingering at the Window of an Inn after Midnight' from *By the Old Walls of Kyoto*, John Weatherhill Inc., 1981. **Randolph Stow:** 'Ishmael', 'The Testament of Tourmaline: Variations on Themes of the *Tao Teh Ching*' from *Selected Poems: A Counterfeit Silence*, Angus & Robertson, a division of HarperCollins Publishers, 1969. **Jennifer Strauss:** 'The Anabaptist Cages, Münster' from *Labour Ward*, Pariah Press, 1988. **Andrew Taylor:** 'That Silence', 'The Invention of Fire', 'The Gods' from *Selected Poems*, University of Queensland Press, 1982. **Dimitris Tsaloumas:** 'Gods' from *The Book of Epigrams*, University of Queensland Press, 1985. **Urumbula People:** 'The Urumbula Song' (translated by T. G. H. Strehlow), published in *Hemisphere* 6:8, 1962. **Vicki Viidikas:** 'Glimpse', 'Varanasi (Uttar Pradesh)', 'Saint and Tomb (Ajmer)' from *India Ink*, Hale & Iremonger, 1984. **Chris Wallace-Crabbe:** 'The Secular' from *Selected Poems*, Angus & Robertson, a division of HarperCollins Publishers, 1973, © Chris Wallace-Crabbe, reproduced by permission of the poet; 'Nor is it Darkness', published in *Eureka Street*, © Chris Wallace-Crabbe, reproduced by permission of the poet; 'Dusky Tracts', © Chris Wallace-Crabbe, reproduced by permission of the poet; 'Sheer Mystery' from *Rungs of Time*, Oxford University Press, 1993. **Maureen Watson:** 'Memo to J. C.', from *Inside Black Australia*, edited by Kevin Gilbert, Penguin Books Australia Ltd, 1988. **Alan Wearne:** 'St Bartholomew Remembers Jesus Christ as an Athlete', published in *Public Relations*, Makar Press, 1972. **Francis Webb:** 'Five Days Old', 'Poet', 'Around Costessey: Good Friday, Norfolk'; 'Back Street in Calcutta', 'Ward Two: Homosexual; A Man', 'St Therese and the Child', 'Lament for St Maria Goretti' from *Cap and Bells: The Poetry of Francis Webb*, Angus & Robertson, a division of HarperCollins Publishers, 1991. **B. R. Whiting:** 'Gandhi, 1946' from *The Poems of B. R. Whiting*, reproduced by permission of The Sheep Meadow Press, 1991. **Judith Wright:** 'Eli, Eli', 'The Lost Man', 'The Harp and the King', 'Vision', 'Prayer', 'For One Dying', 'Eight-panel Screen' from *Collected Poems 1942–1970*, Angus & Robertson, a division of HarperCollins Publishers, 1971; 'Grace' from *Alive*, Angus & Robertson, a division of HarperCollins Publishers, 1973; 'Patterns' from *Phantom Dwelling*, Angus & Robertson, a division of HarperCollins Publishers, 1985. **Yirrkalla People:** 'The Djanggawul Song Cycle: Songs 20–27' from *Djanggawul: An Aboriginal Religious Cult of North-Eastern Arnhem Land* by Ronald M. Berndt, Routledge and Kegan Paul, 1952. **Fay Zwicky:** from 'Ark Voices: Lemur, Whale Psalm' from *Kaddish and Other Poems*, University of Queensland Press, 1982.

Every effort has been made to trace the original source of all copyright material contained in this book. Where the attempt has been unsuccessful, the publisher would be pleased to hear from copyright holders to rectify any errors or omissions.

Index of Themes

Index of First Lines

Index of Poets and Titles